Lecture Notes in Mathematics

Edited by A. Dold and B. Eckmann

1098

T0220032

Groups – Korea 1983

Proceedings of a Conference on Combinatorial Group Theory, held at Kyoungju, Korea, August 26–31, 1983

Edited by A. C. Kim and B. H. Neumann

Springer-Verlag
Berlin Heidelberg New York Tokyo 1984

Editors

Ann Chi Kim
Department of Mathematics, Busan National University
Pusan 607, Republic of Korea

Bernhard H. Neumann
Department of Mathematics, Institute of Advanced Studies
Australian National University
Canberra, ACT 2601, Australia

AMS Subject Classification (1980): 20Exx, 20Fxx, 20Nxx, 05C25

ISBN 3-540-13890-0 Springer-Verlag Berlin Heidelberg New York Tokyo
ISBN 0-387-13890-0 Springer-Verlag New York Heidelberg Berlin Tokyo

Printing and binding: Beltz Offsetdruck, Hemsbach/Bergstr.
2146/3140-543210

PREFACE

The collection of papers here presented came from an international conference on combinatorial theory of groups and related topics held at Kyoungju, Korea, in August 1983. Mrs Barbara M. Geary started the production of the typescript and acted as Editorial Assistant, until ill health forced her to give up the work; Miss Norma Chin then continued with the production to the typescript, and in fact produced almost all of it. Mr Leigh R. Hume then took over the Assistant Editorship and assumed responsibility for the internal proof reading and proof collating. All three are in Canberra, Australia. We here record our gratitude to Mrs Geary, Miss Chin, and Mr Hume. We are also indebted to the Springer-Verlag for being – as always – most cooperative and helpful.

A.C. Kim
B.H. Neumann
Editors

TABLE OF CONTENTS

INTRODUCTION

An international mathematical conference, "Groups — Korea 1983", was held at Kyoungju, Korea, from 26 to 31 August 1983. It was sponsored by the Korea Science and Engineering Foundation (KOSEF), the Korean Educational Ministry, and, most substantially, by its host, Busan National University (BNU). Of the 99 participants, 28 specialists came from outside Korea, and represented 8 countries: Australia, Canada, France, Federal Republic of Germany, Japan, Singapore, the United Kingdom, and the United States of America. The impetus for this conference arose in mathematical correspondence I had, since 1979, with Emeritus Professor Bernhard H. Neumann in Canberra, Australia.

The programme of the conference concentrated on combinatorial group theory and related topics. Ten invited speakers gave one or two one-hour lectures each, in the mornings; a further 13 speakers gave seminar talks of half an hour or an hour each in the afternoons; for the graduate students who attended the conference, there was meanwhile a programme of 14 special invited lectures. The talks given are listed in Appendix A. The participants are listed in Appendix B.

The nonscientific activities of the conference included a Korean traditional meal, a conference dinner, a tour of Kyoungju and surroundings, including a visit to the Kyoungju National Museum, and a climb to the top of Toham Mountain from the Kolon Hotel, which provided both accommodation for the participants and the conference venue. The fact that all participants stayed in the one location made informal contacts easy, and the informal contacts between the graduate students and the overseas experts were perhaps the most stimulating aspect of the conference.

This was the first international mathematical conference held in Korea. It is difficult to estimate its cost. The largest of the direct money contributions was made by the Busan National University, US$12,500; KOSEF contributed US$6,800, the Korean Educational Ministry US$6,250, both to the air fares of overseas delegates; the Bank of Pusan donated US$3,750; the Eun Moon Publishing Co. US$2,500; and a colleague who wishes to remain anonymous US$6,250. Substantial indirect contributions were also made by overseas governments, academies, and scientific foundations, by paying all or part of the air fares of some participants; some universities, both in Korea and elsewhere, similarly contributed to the fares of their members. I take this opportunity of recording my gratitude to all these donors.

It is a pleasure to acknowledge those people who made the conference possible, and who also made it successful. My special thanks go to the then President of Busan National University, Dr Hong-Ju Moon, who, on reaching the compulsory retiring age of 65 years, retired from his official position on the last day of the conference, 31 August 1983. He very much encouraged the organiser of the conference, was always ready to discuss difficult matters of finance, and energetically supported the conference. My thanks also go to all members of the Department of Mathematics of BNU, for their helpful advice; and to Professors Ju-Shil Suh, Dean of Academic Affairs, Tae-Gweon Park, Dean of the College of Liberal Arts, and Hyung-Kyi Lee, Director of the Student Guidance Center, who all helped in the effort of getting financial support for the conference both within and outside the university. Finally, I thank Mr Tae-Ju Park, Director of the Bank of Pusan, for his generous support.

I am much indebted to Professor Bernhard H. Neumann, one of the editors of these Proceedings, for his continuing interest in the conference from its inception, for his invaluable advice, and for numerous helpful comments from the beginning to the end of the conference.

Ann-Chi Kim

Department of Mathematics
Busan National University
6 March 1984

PROCEEDINGS OF 'GROUPS – KOREA 1983'
KYOUNGJU, August 1983

20F10
20F16, 17B30, 16A99

ALGORITHMICALLY INSOLUBLE PROBLEMS ABOUT FINITELY PRESENTED SOLVABLE GROUPS, LIE, AND ASSOCIATIVE ALGEBRAS

Gilbert Baumslag*

1. Introduction

My objective here is to report on some as yet unpublished joint work of Dion Gildenhuys, Ralph Strebel and myself concerned with a variety of algorithmically insoluble problems about finitely presented solvable groups, finitely presented solvable lie algebras and finitely presented associative algebras satisfying a polynomial identity [11], [12], [13] and [14]. The methods used in this work were developed first for groups, where they take on their most complex form. It is therefore fitting that I concentrate first on the results we have obtained for groups. In order to put these results into perspective let me begin by recalling some earlier related work.

2. A little history

In 1955 Novikov [30] (see also [29] and Boone [17]) constructed the first example of a finitely presented group with an insoluble word problem. Shortly afterwards, first Adyan [2], [3], [4] and then Rabin [31], [32] used this group of Novikov to prove that for every one of a large number of group-theoretic properties there is no algorithm whereby one can decide whether any finitely presented group has such a property. In particular there is no algorithm whereby one can decide whether any finitely presented group is trivial, i.e. of order 1 . These negative results of Adyan and Rabin were followed by a number of similar ones involving elements and subgroups of finitely presented groups by Boone, Neumann and myself [7]. For instance we proved that, given an integer $n > 1$, there is no algorithm whereby one can decide whether any element in a finitely presented group is an nth power.

The next 22 years were liberally sprinkled with all kinds of additional negative algorithmic results about finitely presented groups. One of the most striking of these is Miller's proof in 1971 that the isomorphism problem for finitely presented residually finite groups is algorithmically insoluble [26]. None of these theorems

* Support from the N.S.F. is gratefully acknowledged.

applied to finitely presented solvable groups because all of the groups used in the various proofs contained free subgroups of infinite rank.

Then in 1981 Kharlampovich [23], in a remarkable piece of work, settled the last outstanding word problem for groups by proving the

THEOREM 1 (Kharlampovich). *There exists a finitely presented solvable group, of derived length* 3 *, with an insoluble word problem.*

This result raises the possibility that the Adyan-Rabin theorems, among others, can be carried over to finitely presented solvable groups. However it is easy to see that there is an algorithm which decides whether any finitely presented solvable group is trivial. Indeed there is even an algorithm which decides whether a finitely presented solvable group is polycyclic (Baumslag, Cannonito and Miller [8]). So it is not possible to carry over all the negative results for finitely presented groups to finitely presented solvable groups. In fact, as I indicated earlier, there is another difficulty. The methods employed for finitely presented groups do not work for solvable groups even though one can now use Kharlampovich's group instead of Novikov's. I will describe here how some of these difficulties can be overcome.

3. The main theorem

The starting point is the following variation of Kharlampovich's group.

THEOREM 2 (Baumslag, Gildenhuys and Strebel). *There exists a finitely presented solvable group* U *, of derived length* 3 *, with the following properties:*

(i) there is a recursive set $w_1, w_2, \ldots$ *of words in the given generators of* U *such that there is no algorithm whereby one can decide whether or not any of these words take on the value* 1 *in* U *;*

(ii) each of these words $w_1, w_2, \ldots$ *represents an element in the center of* U .

The main difference between Theorem 1 and Theorem 2 is, on the face of it, the condition (ii). Now Kharlampovich's group has trivial center and so it does not satisfy a condition of this kind. In fact it is precisely on this condition that the proofs of most of our applications of Theorem 2 depend. I shall postpone any further comparison between Theorem 1 and Theorem 2 until later. At this point I would prefer to turn to some of the applications of Theorem 2 that I have already alluded to.

4. Some applications of the main theorem

The first application is the negative solution of the isomorphism problem for finitely presented solvable groups.

THEOREM 3 (Baumslag, Gildenhuys and Strebel). *There is a recursive class of finitely presented solvable groups (of derived length* 3 *) such that there is no algorithm whereby one can determine whether or not any pair of groups in the class* *are isomorphic.*

Theorem 3 can be viewed as a generalisation of the negative solution of the isomorphism problem for finitely presented groups as a whole.

Next we have the amusing

THEOREM 4 (Baumslag, Gildenhuys and Strebel). *There is a recursive class of finitely presented solvable groups, each of which is either of derived length* 3 *or of derived length* 4 *, such that there is no algorithm whereby one can decide whether any group in the class is of derived length* 3 .

Thus even the information that a finitely presented group is solvable is insufficient for an algorithmic determination of its derived length. This answers a question raised by Cannonito a few years ago.

Theorems 3 and 4 can be elegantly, albeit easily, deduced from Theorem 2. It is not quite so easy to prove the next one of our applications of the main theorem.

THEOREM 5 (Baumslag, Gildenhuys and Strebel). *There is a recursive class of finitely presented solvable groups of derived length* 3 *such that there is no algorithm whereby one can determine whether any group in the class is directly decomposable.*

Theorems 3, 4 and 5 involve entire classes of groups. The next theorem is concerned with a single group.

THEOREM 6 (Baumslag, Gildenhuys and Strebel). *There is a finitely presented solvable group* G *of derived length* 3 *such that there is no algorithm whereby one can determine whether any word in the given generators represents*

(i) an element of the center of G *;*

(ii) an element that commutes with a given element of G *;*

(iii) an nth *power, where* $n > 1$ *is a given integer;*

(iv) a proper power.

Furthermore, there is no algorithm whereby one can decide whether any given automorphism of G *is*

(v) inner;

(vi) trivial.

As I have already indicated it is the condition (ii) in Theorem 2 that lends itself to the proofs of the theorems that I have been discussing. It is with these proofs that I want to concern myself next, before turning to a discussion of some open problems.

5. The proofs of Theorems 3 and 4

The group U of Theorem 2 is actually one of an infinite family of groups indexed by the set of all primes. Suppose that U is indexed by the prime p. Then, in addition to the properties (i) and (ii) given in Theorem 2, U has the further properties:

(iii) $w_i^p = 1$ $(i = 1, 2, \ldots)$ in U;

(iv) U decomposes into a semidirect product

$$U = P \rtimes A$$

where P is a metabelian group of exponent dividing p^2 and A is torsion-free abelian.

Thus it follows that the p-subgroups of U are all of exponent dividing p^2.

We are now in a position to begin the proof of Theorem 3. Consider then the group

$$Q = \langle a, t; a^{p^3} = 1, t^{p^2} = 1, t^{-1}at = a^{1+p} \rangle .$$

Q is a group of order p^5 with a of order p^3. Notice that

$$b = a^{p^2}$$

is an element of order p in the center of Q.

Now put

$$U_i = (U \times Q)/\mathrm{gp}((w_i, b^{-1})) \qquad (i = 1, 2, \ldots);$$

since w_i is central in U, (w_i, b^{-1}) is central in $U \times Q$ and so $\mathrm{gp}((w_i, b^{-1}))$ is normal in $U \times Q$, i.e. U_i makes sense. If $w_i \neq 1$, then w_i is of order p in U and therefore U_i is the so-called central product of U and Q with w_i identified with b. This implies that both U and Q are embedded in U_i which ensures that U_i contains a p-subgroup that is not of exponent dividing p^2. On the other hand, if $w_i = 1$, then

$$U_i = (U \times Q)/\mathrm{gp}(b) ;$$

so in this case the p-subgroups of U_i are all of exponent dividing p^2. Consequently

$$U_i \cong (U \times Q)/\mathrm{gp}(b) \quad \text{if and only if } w_i = 1 .$$

Thus if there is an algorithm which decides whether any pair of the groups $(U \times Q)/\mathrm{gp}(b)$,

$$U_1, U_2, \ldots$$

are isomorphic, there is an algorithm which decides whether any of the words $w_i = 1$ in U. This proves Theorem 3.

The proof of Theorem 4 is similar to that of Theorem 3. However in place of Q we take the group R of all lower unitriangular matrices of degree 9 over the field F_p of p elements. Then it is not hard to check that R is solvable of derived length 4 and that the third derived group R''' of R is central of order p and it is generated by the element

$$c = I + E_{91} ,$$

with I the identity of R and E_{91} the 9×9 matrix with 1 in the (i,1)th place and 0 everywhere else. We now put

$$V_i = (U \times R)/\mathrm{gp}((w_i, c^{-1})) \quad (i = 1, 2, \ldots) .$$

It follows that V_i is of derived length 4 if $w_i \neq 1$ and is of derived length 3 if $w_i = 1$. This proves Theorem 4.

6. Some open problems

These theorems about finitely presented solvable groups that I have been discussing leave many algorithmic problems, and more generally, problems of a recursive theoretic nature, unresolved. It is to a few of these problems that I want to devote this section.

To begin with we have the

PROBLEM 1. *Is there a finitely presented solvable group with soluble word problem and insoluble conjugacy problem?*

An example of a finitely presented group with soluble word problem and insoluble conjugacy problem was constructed by Fridman [18] in 1960 (see also the elegant examples of Miller [26] in 1971). Neither Theorem 1 nor Theorem 2 seem to shed any light on Problem 1.

Notice that finitely generated metabelian groups are residually finite and hence have solvable word problem (Hall [21]; see also [20]). Indeed, every finitely generated metabelian group has a soluble conjugacy problem (Noskov [28]).

Now it follows from the work of Adyan and Rabin that there is no algorithm which decides whether any finitely presented group is abelian. In other words, the set of all finite presentations of all finitely generated abelian groups is not a recursive subset of the set of all finite presentations. It is, however, a recursively enumerable set. This suggests my next problem.

PROBLEM 2. *Is the set of all finite presentations of all finitely presented solvable groups a recursively enumerable subset of the set of all finite* *presentations?*

The special case of Problem 2 for metabelian groups is itself an interesting problem. Indeed the powerful work of Bieri and Strebel [16] suggests that for metabelian groups the answer to Problem 2 is in the affirmative.

I want to turn next to the integral homology of finitely presented solvable groups. Now the recent joint work of Dyer, Miller and myself [10] contains a partial characterisation of the integral homology of finitely presented groups in terms of recursive function theory. Too little is known about finitely presented solvable groups for there to be a similar characterisation for their integral homology groups. However some information is now available about the integral homology groups of finitely generated metabelian groups (Baumslag, Dyer and Groves [9]). This information, together with the fact that finitely generated metabelian groups have soluble word problems, suggests our next problem.

PROBLEM 3. *Let G be a finitely presented metabelian group, $n > 1$ a positive integer and let H_nG denote the n-dimensional integral homology group of G. Does H_nG have a presentation with soluble word problem?*

In fact I do not know of any example of a finitely presented group with a soluble word problem such that one of its integral homology groups is such that all of its recursively enumerable presentations have insoluble word problems.

Before discussing my last problem, I would like to explicitly describe one of the results contained in this work of Dyer, Groves and myself that I alluded to a few minutes ago: There is an algorithm which decides for every finitely generated metabelian group G whether or not $H_2G = 0$. This theorem should be compared with the corresponding negative theorem of Gordon [19] for the class of all finitely presented groups.

My last concern is with the centers of finitely presented groups. Remeslennikov [33] was the first to construct a finitely presented group with an infinitely generated center. More recently Abels [1] constructed a finitely presented solvable group with an infinitely generated center. Not every countable abelian group can be the center of a finitely presented group. For it is not hard to see that the center of a finitely presented group has a recursively enumerable presentation. This suggests my next problem.

PROBLEM 4. *Can every abelian group with a recursively enumerable presentation be the center of a finitely presented (solvable) group?*

In particular I do not know whether the additive group of rational numbers can be the center of a finitely presented group. Some information about the centers of finitely presented solvable groups has been obtained in unpublished joint work of Thomson and myself [15].

7. A sketch of the proof of the main theorem

The proof of our main theorem, Theorem 2, is based on Kharlampovich's Theorem 1: there exists a finitely presented group, say V, solvable of derived length 3, with an insoluble word problem. Kharlampovich defined her group V in terms of generators and defining relations. She then proved it is solvable of derived length 3. Finally, and this turned out to be the most difficult part of the proof, she proved that V has an insoluble word problem. This part of the proof makes use of the fact that V is an extension of a group of exponent 2 by a metabelian group.

Our group U was obtained from V by omitting two of its generators and then removing its dependence on the prime 2 — thus U is simply one of an infinite family of finitely presented solvable groups of derived length 3 with insoluble word problem. These seemingly minor changes necessitated a completely different approach to the proof. Firstly a description of U in terms of generators and defining relations would have made it too difficult to deal with U. Secondly the technical details of the proof had to be completely redone. The net result was a considerable increase in the clarity of the proof with the added expense of an increase in the length.

In order to explain one of the ideas in Kharlampovich's proof that we made considerable use of let me first remind you of the usual approach to the construction of groups with insoluble word problem. One starts out with a suitably chosen Turing machine and then imprints the workings of that machine into the defining relations of a group. The procedure is as uncomplicated as possible, involving few relations, and so tends to produce lots of free subgroups. Therefore, from the viewpoint of finitely presented solvable groups, it is useless. Kharlampovich's idea was to imprint the workings of a so-called Minsky machine into a finitely presented *solvable* group.

Let me therefore briefly touch on both Turing and Minsky machines.

Turing machines are 1-headed machines. The head of such a machine scans a 2-way infinite tape, can move either to the left or to the right and can both print and erase.

Minsky machines, on the other hand, are 2-headed. Each head scans a 1-way infinite tape which ends on the left. Neither head of the machine can print, while each head can move, independently of the other, either left or right, with leftward movement forbidden once a head scans the end of a tape. Minsky showed that, notwithstanding the restricted nature of his machines, they can be programmed to carry out 4 operations: multiplication by 2, 3 and 5 and, whenever possible, division by 30. This enabled him to prove, rather surprisingly, that given any Turing machine there exists a Minsky machine which essentially duplicates all the computations of the Turing machine itself.

Now given any partial recursive function f there exists a Turing machine which computes $f(n)$ whenever it is defined. This then translates, via Minsky machines, into the observation that every such partial recursive function f can be built up from the four operations multiplication by 2 , 3 , 5 and division by 30 .

In our proof of Theorem 2, the main theorem, we took a graph-theoretical approach to this theorem of Minsky. The starting point is a given partial recursive function f . There then exists a finite graph $\mathcal{F}$, equipped with an orientation, which specifies the way in which f is built up from the four operations that I mentioned a minute ago. The positive edges e of $\mathcal{F}$ are labelled by 6-tuples of the form

$$\langle v, \varepsilon_1, \varepsilon_2, \delta_1, \delta_2, w\rangle ,$$

where v is the origin of e , w its terminus, and

$$\varepsilon_1, \varepsilon_2 \in \{0, 1\} , \qquad \delta_1, \delta_2 \in \{-1, 0, 1\} .$$

Let me say again that this finite oriented graph $\mathcal{F}$ simply describes the manner in which f is built up from the operations multiplication by 2 , 3 , 5 and division by 30 . The aim now is to encode f into a suitably chosen finitely presented solvable group using $\mathcal{F}$. More precisely, guided by Kharlampovich [23], we proceed as follows. First we choose a suitable finitely presented solvable group G — in our case G can be represented as a group of 9×9 matrices over the field of fractions of a polynomial ring in three variables, which makes it relatively easy to work with. We then add finitely many relations to G designed to reflect the graph $\mathcal{F}$. The precise manner in which this is done is not hard to describe. It is with this description that I want to concern myself next.

Let then p be any given prime and let

$$S = F_p[X_1, X_1^{-1}, X_2, X_2^{-1}, (1+X_1)^{-1}, (1+X_2)^{-1}] .$$

We now take H to be the subgroup of $GL(8, S)$ generated by the following 11 triangular matrices, where $\mathrm{diag}(a_1, a_2, \dots, a_8)$ denotes the diagonal matrix with $a_1, a_2, \dots, a_8$ on the main diagonal:

$$x_1 = \mathrm{diag}(1, 1, X_1, 1, X_1, 1, X_1, X_1)$$

$$y_1 = \mathrm{diag}(1, 1, Y_1, 1, Y_1, 1, Y_1, Y_1) \quad (Y_1 = 1 + X_1)$$

$$x_2 = \mathrm{diag}(1, 1, 1, X_2, 1, X_2, X_2, X_2)$$

$$y_2 = \mathrm{diag}(1, 1, 1, Y_2, 1, Y_2, Y_2, Y_2) \quad (Y_2 = 1 + X_2)$$

$$\tilde{x}_i = \mathrm{diag}(1, X_i, 1, 1, X_i, X_i, 1, X_i) \quad (i = 1, 2)$$

$$\tilde{y}_i = \mathrm{diag}(1, Y_i, 1, 1, Y_i, Y_i, 1, Y_i) \quad (i = 1, 2)$$

$$\tilde{d} = 1 + E_{21} + E_{53} + E_{64} + E_{87}$$

$$d_1 = 1 + E_{31} + E_{52} + E_{74} + E_{86}$$

$$d_2 = 1 + E_{41} + E_{62} + E_{73} + E_{85} \, .$$

It follows easily by using the methods introduced in [5] that H is a finitely presented metabelian group.

We need now to construct the group G we mentioned above. To this end let V be the set of all vertices of the finite graph $\mathcal{F}$, which describes the way in which f is built up from the usual operations. We then take M to be the free left S-module with basis

$$\{\beta_{v,k} \mid v \in V , \quad k = 1, 2, \ldots, 8\} \, .$$

We turn M into a right SH-module, where here SH is the group algebra of H over the ring S, in the obvious way:

$$\beta_{v,k} \cdot h = s_{k1}\beta_{v,1} + \cdots + s_{k8}\beta_{v,8} \, ,$$

where, of course,

$$h = (s_{ij})$$

is an 8×8 matrix with entries from S. We have therefore a right action of H on M and can form the semidirect product

$$G = M \rtimes H \, .$$

G is clearly solvable; in fact it is solvable of derived length three. It is easy enough to show that G is also finitely generated. Indeed, but this is not so easy to verify, G turns out to be finitely presented. Incidentally it is easy to show, using the methods introduced by Magnus in the 1930s, that G can be faithfully represented as a group of 9×9 matrices. Thus, in a sense, G is a fairly simple group. On the other hand it is worth noting here that G has a center which, qua vector space over F_p is of infinite dimension and hence is not finitely generated (cf. the discussion preceding Problem 4 in 6).

Our next move is to imprint the graph $\mathcal{F}$ in a quotient group U of G. To this end, let us put

$$t_1 = X_1 + 1 + X_1^{-1} , \quad t_2 = X_2^2 + X_2 \, .$$

Then to each positive edge of $\mathcal{F}$ with label $\langle v, 1, 1, \delta_1, \delta_2, w\rangle$ we assign the element

$$t_1 t_2 \beta_{v,7} - t_1^{1+\delta_1} t_2^{1+\delta_2} \beta_{w,7} \, .$$

To each positive edge of $\mathcal{F}$ with label $\langle v, 1, 0, \delta_1, \delta_2, w\rangle$ we assign the element

$$t_2\beta_{v,4} - t_1^{\delta_1} t_2^{1+\delta_2}\beta_{w,4} .$$

It turns out that there are no positive edges in $\mathcal{F}$ of any other kind. However there are two special vertices in $\mathcal{F}$, a so-called initial vertex α and a so-called final vertex ω , both of which play a special role here. We add, corresponding to this final vertex ω , one more element of G :

$$t_1\beta_{\omega,3} .$$

Now let R be the set of all the elements assigned above to the positive edges of the graph $\mathcal{F}$ and the one to the final vertex ω , and let O be the normal closure in G of R :

$$O = gp_G(R) .$$

Since $\mathcal{F}$ is finite, O is the normal closure of a finite set and therefore

$$U = G/O$$

is a finitely presented solvable group. (Although our notation does not reflect it, U depends on the initial choice of the partial recursive function f .)

Finally we put

$$u_n = t_1^{2^n}\beta_{\omega,3}\, O .$$

Then, and this is the entire point of the procedure that I have outlined,

$$u_n = 1 \text{ in } U \text{ if and only if } n \text{ is in the domain of } f .$$

So if the domain of the partial recursive function f is not recursive, U will be a finitely presented solvable group! Now partial recursive functions of this kind are plentiful (cf. e.g. Maltsev [25]) and the u_n represent elements in the center of U . So this completes the sketch of the proof of Theorem 2.

8. Applications to lie algebras

The approach above can be mimicked for lie algebras. Indeed by making use of the methods in [6] we were able to prove the analogue of Theorem 2 for lie algebras.

THEOREM 7 (Baumslag, Gildenhuys and Strebel). *Given any computable field of characteristic different from two, there exists a finitely presented solvable lie algebra* L *(over this field), of derived length three, with the following properties:*

(i) there is a recursive set $w_1, w_2, \ldots$ *of words in the given generators of* L *such that there is no algorithm which determines whether any one of these words takes on the value* 0 *in* L *;*

(ii) each of the words $w_1, w_2, \ldots$ *represents an element in the center of* L .

This then establishes the existence of a finitely presented solvable lie algebra with an insoluble word problem. In fact the first example of a lie algebra of this kind is due to Kukin [24]. His paper, which has only just appeared, came to our attention some time after our work had been completed.

Theorem 7 has a number of applications to finitely presented solvable lie algebras which are analogous to those of Theorem 2 to finitely presented solvable groups. Thus we have the following theorems.

THEOREM 8 (Baumslag, Gildenhuys and Strebel). *Given any computable field of characteristic different from two, there is a recursive class of finitely presented solvable lie algebras over this field, of derived length three, such that there is no algorithm whereby one can decide whether any pair of lie algebras in the class are isomorphic.*

THEOREM 9 (Baumslag, Gildenhuys and Strebel). *Given any computable field of characteristic different from two, there is a recursive class of finitely presented solvable lie algebras over this field, each of which is either of derived length* 3 *or of derived length* 4 *, such that there is no algorithm whereby one can determine whether any of these lie algebras is of derived length* 3 .

THEOREM 10 (Baumslag, Gildenhuys and Strebel). *Given a computable field of characteristic different from two, there is a finitely presented lie algebra* L *of derived length three over this field with the following properties:*

(i) there is no algorithm whereby one can determine whether any derivation of L *is inner;*

(ii) there is no algorithm whereby one can determine whether any derivation of L *is the zero derivation;*

(iii) there is no algorithm whereby one can determine whether any automorphism of L *is the identity automorphism;*

(iv) there is no algorithm whereby one can determine whether any element of L *is central.*

We have not yet been able to prove the analogue for lie algebras of Theorem 5.

I would like to say at this stage that Kukin's paper [24] contains a wealth of beautiful theorems. Here I want to mention only one consequence of his powerful and important work, which can be formulated as follows.

THEOREM 11 (Kukin). *There exists a computable field and a finitely presented lie algebra over this field such that the word problem is solvable in the lie algebra itself but its universal enveloping algebra has an insoluble word problem.*

Incidentally the methods used by Kukin are very different from ours.

9. Applications to associative algebras satisfying a polynomial identity

The methods developed in the proof of Theorem 2 can be carried over rather easily to associative algebras. In particular we have again the analogue for associative algebras of the corresponding theorem for groups.

THEOREM 12 (Baumslag, Gildenhuys and Strebel). *Given any computable field whatsoever there exists a finitely presented associative algebra* A *satisfying a polynomial identity with the following properties:*

(i) there is a recursive set of words $w_1, w_2, \ldots$ *in the given generators of* A *such that there is no algorithm which determines whether any one of these words takes on the value* 0 *in* A *;*

(ii) each of these words $w_1, w_2, \ldots$ *represents a word in the annihilator of* A.

The usual consequences now follow. Thus there is a recursive class of associative algebras satisfying a polynomial identity with an insoluble isomorphism problem. The obvious analogues of our other theorems also carry over to associative algebras satisfying a polynomial identity.

References

[1] H. Abels, "An example of a finitely presented group", *Homological Group Theory* (London Math. Soc. Lecture Notes Series No. 36. Cambridge University Press, 1979), pp. 205-211, MR82b:20047.

[2] S.I. Adyan, "Algorithmic unsolvability of problems of recognition of certain properties of groups", *Dokl. Akad. Nauk SSSR (N.S.)* 103 (1955), pp. 533-535 (Russian), MR18:455.

[3] S.I. Adyan, "Unsolvability of some algorithmic problems in the theory of groups", *Trudy Moskov. Mat. Obšč.* 6 (1957), pp. 231-298 (Russian), MR20:2370.

[4] S.I. Adyan, "Finitely presented groups and algorithms", *Dokl. Akad. Nauk SSSR (N.S.)* 117 (1957), pp. 9-12 (Russian), MR20:2371.

[5] G. Baumslag, "Subgroups of finitely presented metabelian groups", *J. Australian Math. Soc.* 16 (1973), pp. 98-110, MR48:11324.

[6] G. Baumslag, "Subalgebras of finitely presented solvable lie algebras", *J. of Alg.* 45 (1977), pp. 295-305, MR55:8128.

[7] G. Baumslag, W.W. Boone, and B.H. Neumann, "Some unsolvable problems about elements and subgroups of groups", *Math. Scand.* 7 (1959), pp. 191-201, MR29:1247.

[8] G. Baumslag, Frank B. Cannonito C.F. Miller, "Some recognizable properties of solvable groups", *Math. Z.* 178 (1981), pp. 289-295, MR82k:20061.

[9] G. Baumslag, E. Dyer and J.R.J. Groves, "On the internal homology of finitely generated metabelian groups". In preparation.

[10] G. Baumslag, E. Dyer, and C.F. Miller, "On the integral homology of finitely presented groups", *Topology* 22 (1983), pp. 27-46.

[11] G. Baumslag, D. Gildenhuys, and R. Strebel, "Algorithmically insoluble problems

about finitely presented solvable groups, lie and associative algebras. I". Submitted for publication to *Journal of Pure and Applied Algebra.*

[12] G. Baumslag, D. Gildenhuys and R. Strebel, "Algorithmically insoluble problems about finitely presented solvable groups, lie and associative algebras. II". To appear in *Journal of Algebra.*

[13] G. Baumslag, D. Gildenhuys and R. Strebel, "Algorithmically insoluble problems about finitely presented solvable groups, lie and associative algebras. III". In preparation.

[14] G. Baumslag, D. Gildenhuys and R. Strebel, "Algorithmically insoluble problems about finitely presented solvable groups, lie and associative algebras. IV". In preparation.

[15] G. Baumslag and M.W. Thomson, "On the centres of finitely presented groups". In preparation.

[16] R. Bieri and R. Strebel, "Valuations and finitely presented metabelian groups", *Proc. London Math. Soc.* (3) 41 (1980), pp. 439-464, MR81j:20080.

[17] W.W. Boone, "Certain simple, unsolvable problems of group theory, V., VI", *Nederl. Akad. Wetensch. Proc. Ser. A* 60 = *Indag. Math.* 19 (1957), pp. 22-27, 227-232, MR20:5230.

[18] A.A. Fridman, "On the relation between the word problem and the conjugacy problem in finitely defined groups", *Trudy Moscov. Mat. Obšč.* 9 (1960), pp. 329-356, MR31:1195.

[19] C.McA. Gordon, "Some embedding theorems and undecidability questions for groups". Preprint.

[20] P. Hall, "Finiteness conditions for soluble groups", *Proc. London Math. Soc.* (3) 4 (1954), pp. 419-436, MR17:344.

[21] P. Hall, "On the finiteness of certain soluble groups", *Proc. London Math. Soc.* (3) 9 (1959), pp. 592-622, MR22:1618.

[22] G. Higman, "Subgroups of finitely presented groups", *Proc. Roy. Soc. Ser. A* 262 (1961), pp. 455-475, MR24:A152.

[23] O.G. Kharlampovich, "A finitely presented solvable group with insoluble equality problem", *Izv. Akad. Nauk Ser. Mat.* 45 (1981), no. 4, pp. 852-873 (Russian), MR82m:20036.

[24] G.P. Kukin, "The equality problem and free products of lie algebras and of associative algebras", *Sib. Mat. Ž.* 24 (1983), no. 2, pp. 85-96 (Russian).

[25] A.I. Maltsev, *Algorithms and recursive functions* (English translation, Walters-Noordhoof, Groningen, Netherlands, 1970), MR41:8233.

[26] C.F. Miller, *On group-theoretic decision problems and their classification* (Ann. Math. Studies 68, Princeton University Press, 1971), MR46:9147.

[27] M.L. Minsky, "Recursive unsolvability of Post's problem of "tag" and other topics in the theory of Turing machines", *Ann. of Math.* 74 (1961), pp. 437-455, MR25:3825.

[28] G.A. Noskov, "On the conjugacy problem in metabelian groups", *Mat. Zametki* 31 (1982), no. 4, pp. 495-507 (Russian), MR83i:20029.

[29] P.S. Novikov, "On algorithmic unsolvability of the problem of identity", *Dokl.*

Akad. Nauk SSSR (N.S.) 85 (1952), pp. 709-712 (Russian), MR14:618.

[30] P.S. Novikov, *On the algorithmic unsolvability of the word problem in group theory* (Trudy Mat. Inst. im. Steklov. No. 44, Izdat. Akad. Nauk, SSSR, Moscow, 1955) (Russian), MR17:706.

[31] Michael O. Rabin, "Recursive unsolvability of group theoretic problems", *Bull. Amer. Math. Soc.* 62 (1956), p. 396.

[32] Michael O. Rabin, "Recursive unsolvability of group theoretic problems", *Ann. of Math.* (2) 67 (1958), pp. 172-194, MR22:1611.

[33] V.N. Remeslennikov, "A finitely presented group whose center is not finitely generated", *Alg. i Logika* 13 (1974), no. 4, pp. 450-459 (Russian), MR52:14064.

The City College of CUNY
Convent Av. at 138 Street
New York, N.Y. 10031

PROCEEDINGS OF 'GROUPS — KOREA 1983'
KYOUNGJU, August 1983 20D06

ON THE SIMPLE GROUPS OF ORDER LESS THAN 10^5

Colin M. Campbell and Edmund F. Robertson

1. Introduction

The non-abelian simple groups G with $|G| < 10^6$, excluding the groups $PSL(2, p^n)$, have been studied in a series of papers. Maximal subgroups and character tables are given in [6] and [8] respectively. All minimal generating pairs for these groups are given as permutations of minimal degree d in [9]. The pair (a, b) is a *minimal generating pair* for G if $G = \langle a, b\rangle$, a has minimal order among those elements which, together with one other element, generate G and b has minimal order among those elements which together with the fixed element a generate G. It is a consequence of [9] that a is an involution for the non-abelian simple groups of order $< 10^6$. Presentations on all the minimal generating pairs are given in two papers, [5] giving the presentations for the groups of order $< 10^5$ while [4] gives the presentations for those G with $10^5 < |G| < 10^6$. An attempt to find presentations with a minimal number of relations is discussed in [2]. The papers [4] and [5] also give, for each minimal generating pair, words in these generators which generate a subgroup H of minimal index d in G.

In this paper the techniques and computer programs developed to obtain the results in [4] are applied to obtain further information about the 13 groups G with $|G| < 10^5$. In particular, for each minimal generating pair we obtain exactly two words $x = w_1(a, b)$, $y = w_2(a, b)$ and $H = \langle x, y\rangle$ and we give a presentation for H on the generators x and y. Note that for PSU(4, 2), Sz(8) and PSU(3, 4), [5] gives three and four subgroup generators in certain cases.

We also show that a number of the presentations given in [5] contain redundant relations. Omitting these redundant relations, no presentation on a minimal generating pair for these groups of order less than 10^5 contains more than 6 relations.

In addition to the computer programs referred to above, use was also made of a Todd-Coxeter coset enumeration program, a Reidemeister-Schreier program, a Tietze transformation program [7] and a modified Todd-Coxeter program [1]. The first two of these programs, namely Todd-Coxeter and Reidemeister-Schreier, are based on programs originally developed in the Australian National University.

We shall use the notation $[r, s] = r^{-1}s^{-1}rs$ and $r^s = s^{-1}rs$. The minimal permutation generators of [9] and the presentation which they satisfy in [5] are given the same number and we shall use the numbering system as in these papers. For example PSL(3, 3) has two minimal generating pairs (up to action by automorphisms) and the minimal generating pairs of permutations (a, b) are given in 5.1 and 5.2 of [9] while presentations satisfied by these generating pairs are given under 5.1 and 5.2 of [5].

2. The groups G, $|G| < 10^4$

The groups A_5, PSL(3, 2) and A_6 have only one minimal generating pair. The subgroup generators $x = a^b$ and $y = b^a$ generate A_4 of minimal index $d = 5$ in A_5, where (a, b) is the minimal generating pair 1.1, and satisfy the presentation

$$A_4 = \langle x, y \mid x^2 = y^3 = (xy)^3 = 1\rangle .$$

Similarly, with (a, b) as in 2.1, $x = a^{b^2a}$, $y = b$ generate S_4 of minimal index $d = 7$ in PSL(3, 2) and satisfy

$$S_4 = \langle x, y \mid x^2 = y^3 = (xy)^4 = 1\rangle .$$

For A_6 and A_7 presentations for subgroups of minimal index on the given subgroup generators are given in [5].

For PSL(3, 3) there are two minimal generating pairs. Taking $x = a^{ba}$, $y = b$ in 5.1 and $x = a^{ba}$, $y = b^{ab}$ in 5.2 we obtain a subgroup of minimal index $d = 13$, the Hessian group of order 216 extended by C_2, each pair satisfying the presentation

$$\langle x, y \mid x^2 = y^3 = (xy)^8 = ((xy)^2xy^{-1})^2(xy(xy^{-1})^2)^2 = 1\rangle .$$

It is worth noting that PSL(3, 3) is presented in the following symmetric way in [3]:

$$\mathrm{PSL}(3, 3) = \langle r, s \mid r^{13} = 1, r^4sr = s^3, s^4rs = r^3\rangle .$$

Now $a = sr^3s^{-2}r^2$ and $b = rs$ are a minimal generating pair for PSL(3, 3) and satisfy 5.1 of [5]. The information given above allows one to compute, in the symmetric presentation above, generators for a subgroup of minimal index.

For PSU(3, 3) there are again two minimal generating pairs. Taking $x = a^b$ and $y = (ab)^3b^3$ in 6.1 and $x = a$, $y = a^bb$ in 6.2 we obtain a subgroup of minimal index $d = 28$ with presentation

$$\langle x, y \mid x^2 = y^8 = (xy^4)^3 = xyxy^{-1}xy^{-2}xy^2 = 1\rangle .$$

This subgroup, of order 216, cannot be the Hessian group as stated in [6] since its derived factor group is C_8 and its derived group is the Burnside group B(2, 3).

We note also that the presentation 6.2 of [5] contains redundant relators. Two possible subsets of the relations which suffice to define PSU(3, 3) are contained in the following presentations:

$$\mathrm{PSU}(3,3) = \langle a, b \mid a^2 = b^6 = (ab)^8 = [a, b]^4 = (ab^3)^3 = ((ab)^2ab^2)^3 = 1\rangle$$

$$= \langle a, b \mid a^2 = (ab)^8 = (ab^3)^3 = (ab^2)^3(ab^{-2})^3 = ((ab)^2ab^2)^3 = 1\rangle .$$

M_{11} has two minimal generating pairs but, denoting the pair in 7.1 by (a, b), the pair of 7.2 is (a, b^{-1}) so only the first pair need to be studied. The relation $b^4 = 1$ is redundant in presentation 7.1 of [5] so we have

$$M_{11} = \langle a, b \mid a^2 = (ab)^{11} = (ab^2)^6 = (ab)^2(ab^{-1})^2abab^{-1}ab^2abab^{-1} = 1\rangle .$$

A subgroup H of minimal index 1 is M_{10} generated by $x = a$, $y = b^{(ab)^2}$ with presentation

$$M_{10} = \langle x, y \mid x^2 = y^4 = (xy^2)^5 = (xy^{-1}xy^2)^4y^{-1}xy = 1\rangle .$$

3. The groups G, $10^4 < |G| < 10^5$

For A_8 a subgroup of minimal index $d = 8$ is given by $\langle b^2, (bab)^{aba}\rangle$ for 8.1 and $\langle a, (ab^2)^2ab\rangle$ for 8.2. These subgroup generators satisfy the presentations 4.1 and 4.2 for A_7 respectively.

There is only one minimal generating pair for PSL(3, 4). A subgroup H of minimal index $d = 21$ is generated by $x = aba$, $y = a^{bab^2}ab$. A presentation for H on x and y is

$$\langle x, y \mid x^4 = y^5 = (xy)^3 = (xy^{-1})^3 = (x^2y)^5 = (x^2y^2xy^{-2})^2 = 1\rangle .$$

This group is an extension of an elementary abelian 2-group of order 16 (generated by x^2, $y^{-1}x^2y$, $y^{-2}x^2y^2$, $y^{-3}x^2y^3$) by A_5. No proper normal subgroup contains an element of order 4 so the information given in [6] concerning this subgroup of minimal index cannot be correct. Note that 9.1 of [5] may be replaced by the neater presentation

$$\langle a, b \mid a^2 = b^4 = (ab)^7 = (ab^2)^5 = ((ab)^3b^2)^5 = ((ab)^3bab^{-1})^5 = 1\rangle .$$

For PSU(4, 2) there are 9 minimal generating pairs and we present the information in tabular form. Note that in each case we find just two subgroup generators x and y for a subgroup H of minimal index d and in each case a presentation for H on x and y is

$$\langle x, y \mid x^2 = y^5 = (xy^2)^5 = (xyxy^2)^4 = (xyxyxy^{-1}xy^2)^2 = 1\rangle .$$

The notation $b \to b^{-1}$ means that if the preceding minimal generating pair for the group was (a, b) the generating pair now considered is (a, b^{-1}) . The column headed (†) notes redundant relations in the presentation given in [5], so that in the presentation 10.3, for example, the fourth relation may be omitted.

PSU(4, 2) order = 25920 = $2^6.3^4.5$ $d = 27$

		x	y	(†)
10.1		a	$b^2ab(babab)^2$	-
10.2		a	$b^2(b^{-1}ab^2a)^2$	-
10.3		a	$bab^{-1}(ab)^3b$	4
10.4	$b \to b^{-1}$			
10.5		a	$b^2aba.a^{bab}$	3 & 4
10.6	$b \to b^{-1}$			
10.7		b^2	$ab^{-1}(ab)^4b^2$	-
10.8		$(ba)^2(b^{-1}a)^2$	b^3	-
10.9		$ab^{-2}ab^2abab^{-1}$	b	3

Note that in 10.7 of [5] the fourth relation may be replaced by the neater relation $(abab^{-1}ab^2)^5 = 1$.

For Sz(8) there are 16 minimal generating pairs and again we present the information as a table.

A presentation for a subgroup H of minimal index $d = 65$ on generators r and s is

$$\langle r, s \mid r^4 = s^7 = (rs^{-2})^7 = [r^2, srs^{-1}] = rsr^{-1}s^{-3}rs^2 = 1\rangle .$$

We give, in each case, a pair of generators (x, y) for H and its relation to the generating pair (r, s) .

There are two minimal generating pairs for PSU(3, 4) . Taking $r = b$, $s = (ab(ab^{-1})^2)^2ab^{-1}aba$ in 12.1 then $\langle r, s\rangle$ is a subgroup of minimal index $d = 65$ with presentation

$$\langle r, s \mid (rs)^2s(r^{-1}s)^2 = rsr^2s^{-1}rs^{-4}r^{-1}s^{-1} = 1\rangle . \qquad (*)$$

Taking $x = (ab)^5a$, $y = bab^{-1}(ab)^6$ in 12.2 we obtain a subgroup of minimal index with presentation

$$\langle x, y \mid x^2y^4xy^{-1} = (x^2y)^2xy^{-2} = 1\rangle . \qquad (**)$$

The presentations (*) and (**) are related by the simple change of generating pairs $(x, y) = (s, s^{-1}r^{-1})$.

It is interesting to note that these are deficiency zero presentations for this subgroup of order 960 .

Sz(8) order = 29120 = $2^6.5.7.13$ $d = 65$

		x	y	(r, s)	(†)
11.1		b^{ab}	$(ba)^2(b^2a)^2$	(x, y)	-
11.2	$b \to b^{-1}$				
11.3		b^a	$b^{-1}(b^2)^{ab^{-1}ab}$	(x, y)	-
11.4	$b \to b^{-1}$				
11.5		ab	$(ab^{-1})^2(ab)^2ab^{-1}$	(y, x)	-
11.6	$b \to b^{-1}$				
11.7		b^a	$babab^2$	(x, xy)	-
11.8	$b \to b^{-1}$				
11.9		b^a	$bab^{-1}ab^2ab$	(x, y^{-1})	6
11.10	$b \to b^{-1}$				
11.11		$b^2ab^{-1}(ab)^2$	$(ab)^{ba}$	(x, xy)	-
11.12	$b \to b^{-1}$				
11.13		b^a	$ba(bab)^3$	(x, xy)	6
11.14	$b \to b^{-1}$				
11.15		b^a	$(ba)^2b^2ab$	(x, y^{-2})	3
11.16	$b \to b^{-1}$				

Finally for M_{12} there are three minimal generating pairs. In 13.1 the subgroup generators $x = a$ and $y = ab^{-1}(ab^{-1}(ab)^3)^2ab$ generate M_{11} and satisfy 7.1; in 13.2 the subgroup generators $x = a$ and $y = bab^{-1}a(b^{-1}aba)^2$ generate M_{11} and satisfy 7.2; in 13.3 the subgroup generators $x = b^{(ab^{-1}ab)^2abab}$ and $y = a^{bab}(ab^{-1})^3$ generate M_{11} and satisfy 7.1. The presentations 13.2 and 13.3 of [5] both contain a redundant relation so we now have the simpler presentations for 13.2

$$\langle a, b \mid a^2 = b^3 = (ab)^{11} = [a, b]^6 = (ababab^{-1})^6 = 1\rangle$$

and for 13.3

$$\langle a, b \mid a^2 = b^3 = (ab)^{11} = [a, b]^6 = ((ab)^4ab^{-1}abab^{-1})^3 = 1\rangle$$

References

[1] D.G. Arrell and E.F. Robertson, "A modified Todd-Coxeter algorithm", *Proc. L.M.S. Durham Conf. on Computational Group Theory*, to appear.

[2] C.M. Campbell and E.F. Robertson, "The efficiency of simple groups of order $< 10^5$", *Comm. Alg.* 10 (1982), pp. 217-225, Zbl.478.20024.

[3] C.M. Campbell and E.F. Robertson, "Some problems in group presentations", *J. Korean Math. Soc.* 19 (1983), pp. 59-64.

[4] C.M. Campbell and E.F. Robertson, "Presentations for the simple groups G,

$10^5 < |G| < 10^6$", *Comm. Alg.*, to appear.

[5] J.J. Cannon, J. McKay, and K.-C. Young, "The non-abelian simple groups G, $|G| < 10^5$ – presentations", *Comm. Alg.* 7 (1979), pp. 1397-1406, MR80e:20023.

[6] J. Fischer J. McKay, "The non-abelian simple groups G, $|G| < 10^6$ – maximal subgroups", *Math. Comp.* 32 (1978), pp. 1293-1302, MR58:16867.

[7] G. Havas, P.E. Kenne, J.S. Richardson, and E.F. Robertson, "A Tietze transformation program", *Proc. L.M.S. Durham Conf. on Computational Group Theory*, to appear.

[8] J. McKay, "The non-abelian simple groups G, $|G| < 10^6$ – character tables", *Comm. Alg.* 7 (1979), pp. 1407-1445, MR80e:20024.

[9] J. McKay K.-C. Young, "The non-abelian simple groups G, $|G| < 10^6$ – minimal generating pairs", *Math. Comp.* 33 (1979), pp. 812-814, MR80d:20018.

Mathematical Institute
University of St Andrews
St Andrews, Fife, Scotland

PROCEEDINGS OF 'GROUPS — KOREA 1983'
KYOUNGJU, August 1983

20F10
20F16, 17B10

ON SOME ALGORITHMIC PROBLEMS FOR FINITELY PRESENTED GROUPS AND LIE ALGEBRAS

Frank B. Cannonito

1. Introduction

The purpose of this work is to present a collection of algorithmic problems for finitely presented groups and Lie algebras together with some indication of their origin and current status. While most of the problems described will be concerned with finitely presented solvable groups, two, due to Higman are not. They will be discussed together with an hitherto unperceived connection between them.

2. Finitely presented solvable groups

Here there are two basic themes which present themselves. First, the search for analogs of Higman's Embedding Theorem [12] in certain varieties of solvable groups, and second, the question of whether a finitely presented solvable group of a certain type has solvable word problem.

2.1 Varietal analogs of Higman's Theorem

The idea of a varietal analog of Higman's Embedding Theorem was first suggested by Baumslag [2] and further discussed by Cannonito [8]. I will review the details. Baumslag proved a finitely generated metabelian group can be embedded in a finitely presented metabelian group [2]. Thus, it is possible to view this result as an analog of Higman's Embedding Theorem for the variety $\mathfrak{A}^2$ of solvable groups of derived length at most 2 . Baumslag then inquired if further analogs in varieties of Higman's Theorem were to be found. For several years it was erroneously believed that for the solvable varieties, at least, the answer was no, usually on cardinality grounds. For example, since there was a continuum of nonisomorphic finitely generated center-by-metabelian groups it was held by some that no hope of an analog of Higman's embedding theorem for the variety $\mathfrak{C}$ of center-by-metabelian groups was possible on this account. Of course, the same would be true for any variety containing $\mathfrak{C}$ such as $\mathfrak{N}_c\mathfrak{A}$, the variety of nilpotent of class c-by-abelian groups, $c \geq 2$.

As discussed by Cannonito [8] the notion of just how to frame an analog of Higman's embedding theorem varies from variety to variety. The reason it was held that no analog of Higman's theorem was possible for the variety center-by-metabelian was due to a faulty understanding of how to pose the analog. In fact when the analog was properly posed, it turned out that the analog was shown by Strebel to exist [16]. That is, Cannonito argued that the conditions which are locally inherited of finitely presented center-by-metabelian groups cut down the size of the class of finitely generated center-by-metabelian groups that one must seek to embed. Therefore, an analog of Higman's embedding theorem for a variety should be a characterization of the finitely generated subgroups of the finitely presented groups of the variety. This led to the following theorem, first conjectured in [8].

THEOREM (Strebel [16]). *A necessary and sufficient condition that a finitely generated center-by-metabelian group be embeddable in a finitely presented center-by-metabelian group is that it be abelian-by-polycyclic.*

The conjecture and theorem above were found by a procedure described in [8]. Namely, one studies the local properties of the finitely presented groups in question and seeks only to embed finitely generated groups satisfying those properties. Thus, the finitely presented center-by-metabelian groups were shown by Groves [10] to be abelian-by-polycyclic, and since such groups are residually finite they have solvable word problem. Now in general, it is not true that a finitely generated residually finite group will have solvable word problem even if recursively presented (see, for example [15], p.196). However for finitely generated abelian-by-polycyclic groups solvability of the word problem follows, for example, from the property max-n and Baumslag, Cannonito and Miller [3], Theorem 2.2. Thus, in framing the conjecture the condition abelian-by-polycyclic subsumed the locally inherited property of having solvable word problem. This is not the case, however, for the variety $\mathbf{N}_2\mathbf{A}$ which we now discuss.

The most tempting variety to study presently is $\mathbf{N}_2\mathbf{A}$: nilpotent of class 2 by abelian. Here it is known from Bieri-Strebel [5] that the finitely presented groups in $\mathbf{N}_2\mathbf{A}$ have max-n , are residually finite and therefore have solvable word problem. It is also known that if G is such a group, then the second derived group G'' has boundedness of exponent on its torsion subgroup. All of the above properties are inherited by subgroups.

Here are some examples due to Strebel (private communication) to illustrate these points. We start with

$$\tilde{G} = \left\{ \begin{pmatrix} 1 & * & * \\ & X^j & * \\ & & 1 \end{pmatrix} : j \in \mathbb{Z} \ , \ * \in \mathbb{Z}[X, X^{-1}] \right\} \ ,$$

where X is an indeterminate. $\tilde{G}$ is generated by the three elements

$$\mathrm{diag}(1, X, 1)\ , \quad \begin{pmatrix} 1 & 1 & 0 \\ & 1 & 0 \\ & & 1 \end{pmatrix}, \quad \text{and} \quad \begin{pmatrix} 1 & 0 & 0 \\ & 1 & 1 \\ & & 1 \end{pmatrix}$$

and its center is isomorphic to the additive group of $Z[X, X^{-1}]$ which is freely generated by the monomials. Let $p > 1$ be a natural number and define

$$Z = \langle p^{|j|} X^j : j \in \mathbb{Z} \rangle$$

and put $G_1 = \tilde{G}/Z$. Then G_1 is a finitely generated center-by-metabelian group; its center has unbounded torsion but G is residually finite.

Next, let A be a recursively enumerable subset of the natural numbers and set

$$Z^+(A) = \langle X^0,\ X^{2^j} : j \in A \rangle .$$

Then $G_3^+(A) = \tilde{G}/Z^+(A)$ is torsion-free and has solvable word problem if and only if A is recursive.

Finally, reduce the ring $\mathrm{mod}(X^{2^k} - 1)$, where $k \geq 0$. Then

$$\mathbb{Z}[X, X^{-1}]/Z^+(A) + (X^{2^k} - 1).\mathbb{Z}X$$

$$\cong (\mathbb{Z}X^0 \oplus \mathbb{Z}X^1 \oplus \ldots \oplus \mathbb{Z}X^{2^{k-1}})/(\mathbb{Z}X^0 + \Sigma\mathbb{Z}(X^{2^j} : j \in A,\ j < k)$$

$$\cong \oplus\{\mathbb{Z}X^\nu : 1 \leq \nu \leq 2^k\ ,\ \nu \neq 2^j \quad \text{with} \quad j \in A\} .$$

Then

$$G_2(A) = \left\{ \begin{pmatrix} 1 & * & * \\ & X^j & * \\ & & 1 \end{pmatrix} : * \in \mathbb{Z}[X, X^{-1}] \right\} \Big/ \left\langle \begin{pmatrix} 1 & 0 & p^j X^j \\ & 1 & 0 \\ & & 1 \end{pmatrix} : j \in A \right\rangle$$

has unbounded torsion, solvable word problem if and only if A is recursive, and is residually finite. By not tossing X^0 in above, one obtains a group $G_3^-(A)$ which is not residually finite. This follows because in every finite homomorphic image of $\tilde{G}/Z^-$, where $Z^- = \Sigma\{ZX^j : j = 2^k,\ k \in A\}$ the matrix $\mathrm{diag}(1, X, 1)$ must have finite order, say $x^n = 1$ where $x = \mathrm{diag}(1, X, 1)$. So some reduction relation $X^n \equiv 1$ is unavoidable. If we now arrange that A has all the even numbers then the quotient

$$\mathbb{Z}X/(X^n - 1)\mathbb{Z}X + Z^-$$

is always a homomorphic image of

$$\mathbb{Z}X/(X^n - 1)\mathbb{Z}X + \Sigma\{\mathbb{Z}X^{2^{(2k)}} : k \in \mathbb{N}\} ,$$

as the congruences $4^k \equiv 1 \pmod{n}$ always have a solution for $k \geq 0$, namely the order of 4 in $\mathbb{Z}/n\mathbb{Z}$. We conclude that

$$\begin{pmatrix} 1 & 0 & 1 \\ & 1 & 0 \\ & & 1 \end{pmatrix} . Z^{-} \in G_3^{-}(A)$$

is in the kernel of every map onto a finite group. Similarly, we can modify G_2 to obtain examples which are not residuably finite. In summary, we can construct examples of center-by-metabelian groups which are finitely generated recursively presented and satisfy any subset of the following three properties:

(i) residual finiteness,

(ii) boundedness of torsion,

(iii) solvable word problem.

Thus, properties (i), (ii) and (iii) are independent and the conjectured Higman embedding theorem for the variety $\mathbf{N}_2\mathbf{A}$ takes the following form.

CONJECTURE. *A necessary and sufficient condition for a finitely generated nilpotent of class 2 by abelian group G to be embeddable in a finitely presented nilpotent of class 2 by abelian group is that G satisfy properties (i) and (iii) above and that G'', the second derived group of G, have boundedness of torsion.*

2.2 The word problem for finitely presented solvable groups

A problem which is a companion to the search for analogs of Higman's Embedding Theorem is that of determining necessary and sufficient conditions for a finitely presented solvable group to have solvable word problem. Indeed, here we will see the most interesting question for the moment concerns the role of the condition max-n. I will review the known facts.

In each of the varieties of solvable groups considered thus far, viz. $\mathbf{A}^2$ (metabelian), $\mathbf{C}$ (center-by-metabelian) and $\mathbf{N}_2\mathbf{A}$ (nilpotent of class 2-by-abelian) the finitely presented groups have solvable word problem. Essentially, this behavior is subsumed by the following, more general.

THEOREM (Baumslag, Cannonito and Miller [3]). *A nilpotent-by-polycyclic group with* max-n *has solvable word problem.*

Recently Harlampovich exhibited a finitely presented group G with unsolvable word problem in the class $\mathbf{N}_4\mathbf{A} \cap \mathbf{A}_2\mathbf{A}_2\mathbf{A}_1$ where $\mathbf{A}_2$ is the class of abelian groups of exponent 2 [11]. This result may be regarded as the crowning achievement in the seventy year long saga of the interaction of the two seemingly dissociated branches of mathematics, mathematical logic and group theory. Beginning in 1911 with the framing by Dehn of the word problem (and by his continued emphasis on its importance) and in the mid-fifties by the independent discoveries of Novikov and Boone of finitely presented groups with unsolvable word problem and culminating in the work of Harlampovich, the contribution of mathematical logic to pure group theory has been decisive, and I have yet to mention the fundamental result of Higman referred to above, also made possible by the contribution from mathematical logic of the precise definition of an algorithm. Therefore, what remains now to be done in this subject includes the determination of the precise boundary where solvable word problem passes into the unsolvable. Thus, we have the following situation. The finitely presented groups in $\mathbf{N}_2\mathbf{A}$ have solvable word problem, the finitely presented groups in $\mathbf{N}_4\mathbf{A}$ may have unsolvable word problem. What is the situation for the finitely presented groups in $\mathbf{N}_3\mathbf{A}$?

Here is what is known to me. The example of Abels [1] shows that a finitely presented group in $\mathbf{N}_3\mathbf{A}$ need not satisfy max-*n* . Thus, the theorem of Baumslag, Cannonito and Miller [3] will not bear on the ultimate disposition of this problem. Of course, Abel's group, being residually finite (as a linear group) has solvable word problem. Since Abel's group fails to have max-*n* , this property for the classes $\mathbf{N}_c\mathbf{A}$, $c \geq 2$ is only sufficient for solvability of the word problem, but not necessary.

The only necessary and sufficient condition for a class of solvable groups to have solvable word problem known to me is the following.

THEOREM (Cannonito and Robinson [9]). *A necessary and sufficient condition for a finitely generated solvable group of finite (Prüfer) rank to have solvable word problem is that it be recursively presentable.*

Recall a group is said to have *finite rank* if there is a finite upper bound to the minimal number of generators needed to generate a finitely generated subgroup. The result above has the immediate corollary.

COROLLARY. *A finitely presented solvable group of finite rank has solvable word problem.*

Here the property max-*n* if included, completely trivializes the problem of solvability of the word problem because solvable groups of finite rank with max-*n* are nilpotent by abelian by finite and the theorem of Baumslag, Cannonito and Miller [3] applies.

Thus, I wish to record at this point the following two problems which appear to me to be the most interesting in this area.

PROBLEM 1. *Does a finitely presented* $\mathbf{N}_3\mathbf{A}$ *group necessarily have solvable word problem?*

PROBLEM 2. *Does a finitely presented solvable group with* max-*n* *necessarily have solvable word problem?*

Derek Robinson has constructed a finitely generated but not recursively presented solvable group of derived length 3 and satisfying max-*n* which fails to have solvable word problem.

3. Two problems of Higman

The first example of a finitely presented infinite simple group was constructed by Thompson (unpublished). Here and henceforth we do not include the trivial group among simple groups. Building on this example, Higman constructed a recursively enumerable sequence of finitely presented infinite simple groups [13]. Naturally the question arises as to whether or not there are other finitely presented infinite simple groups. Bearing on this is

PROBLEM 3 (Higman). *Is the class of finitely presented simple groups recursively enumerable?*

Note that a negative answer to Problem 3 implies there are indeed more groups than those arising in Higman's sequence. At this point, I want to exhibit an algorithm which solves the word problem in a finitely presented simple group. This algorithm (or something like it) was first noted by Kuznetsov in 1958. Let G be a finitely presented simple group with presentation

$$\langle a_1, a_2, \ldots, a_n; R_1, R_2, \ldots, R_m\rangle .$$

Here, R_i is a word on the a's. It is well-known that to solve the word problem for such a presentation one must merely give a recursive enumeration of the words W unequal to 1. Let W be an arbitrary word on the a's. We start two potentially infinite searches going. For the first we recursively enumerate all the consequences of the $R_1, \ldots, R_m$. For the second, we recursively enumerate all the consequences of the $R_1, \ldots, R_m$ together with W. Now, if W equals 1 in G the word W will turn up on the first list in finitely many steps. If W is unequal to 1 in G then adding W to the relators $R_1, \ldots, R_m$ results in a presentation of the trivial group since G is known to be simple. Thus, the generators a_i of which there are finitely many, must all turn up in the second list in finitely many steps. In this way we can determine if W is unequal to 1 in G and thereby recursively enumerate the words W unequal to 1 solving the word problem thereby. Note that this procedure is uniform in the given data, that is, there is a uniform way to solve the word problem in finitely presented simple groups. (The reason we have excluded the trivial group from the class of simple groups is now apparent; the algorithm above does not work for trivial groups.) We are now ready to discuss the second problem.

In 1973 Boone and Higman gave a succinct algebraic characterization of finitely generated groups with solvable word problem. We state the latest refinement of this theorem in the following form.

THEOREM (Thompson [17]). *A necessary and sufficient condition for a finitely generated group G to have solvable word problem is that there exist a finitely generated recursively presentable simple group H containing G as a subgroup.*

The most interesting question yet remaining here is whether or not H can be assumed finitely presentable. If it turns out that it is not possible to so assume, we will say the Boone-Higman theorem is *the best possible* (theorem, that is). This leads to

PROBLEM 4 (Higman). *Show the Boone-Higman theorem is the best possible.*

We now observe that a positive solution to Problem 3 of Higman implies a positive solution to Problem 4 of Higman, namely if the class of finitely presented simple groups is recursively enumerable and if every finitely generated group with solvable word problem is embeddable in a finitely presented simple group, there would be a uniform method for solving the word problem for all finitely generated groups with solvable word problem, in violation of the theorem of Boone-Rogers [7]. Thus, Problems 3 and 4 of Higman are not independent, as formerly believed.

4. Connections with Lie algebras

Most of the ideas presented above go over into comparable questions about Lie algebras which, for definiteness, we take over the rationals.

Thus, for example, the analog of Baumslag's metabelian embedding theorem has an exact counterpart in Lie algebras. Such is also the case for the result of Harlampovich. Namely it was shown by Kukin [14] and independently by Baumslag, Guildenhuys and Strebel [4], there exists a finitely presented solvable Lie algebra of derived length three with unsolvable word problem. While I have not checked the details, it seems likely, e.g., that we can prove there is an algorithm which can decide if an arbitrary finitely presented solvable Lie algebra is finite dimensional. This result is the analog of Theorem 3.3 of [3] with "finite dimensionality" playing the role of polycyclicity. For more in connection with this, see the paper of Bokut' [6].

References

[1] H. Abels, "An example of a finitely presented soluble group", *Homological Group Theory* (London Math. Soc. Lecture Notes, Series **36**, Cambridge Univ. Press, 1979), pp. 205-211, MR82b:20047.

[2] G. Baumslag, "On finitely presented metabelian groups", *Bull. A.M.S.* 78 (1972), p. 279, MR45:354.

[3] G. Baumslag, F.B. Cannonito, and C.F. Miller, III, "Some recognizable properties of solvable groups", *Math. Z.* 178 (1981), pp. 289-295, MR82k:20061.

[4] G. Baumslag, D. Guildenhuys, and R. Strebel, "Algorithmically insoluble problems about finitely presented solvable groups and Lie algebras". To appear.

[5] R. Bieri and R. Strebel, "Valuations and finitely presented metabelian groups", *Proc. London Math. Soc.* (3) **41** (1980), pp. 439-464.

[6] L.A. Bokut', "New results in the theory of associative and Lie rings", *Algebra i*

Logika 20 (1981), no. 5, pp. 531-545 (Russian), Zbl.488.17001, 496.17001.

[7] W.W. Boone and H. Rogers, Jr., "On a problem of J.H.C. Whitehead and a problem of Alonzo Church", *Math. Scand.* 19 (1966), pp. 185-192, MR35:1465.

[8] F.B. Cannonito, "On varietal analogs of Higman's embedding theorem", to appear in *The Festschrift for Roger Lyndon*, AMS.

[9] F.B. Cannonito and D.J.S. Robinson, "The word problem for finitely generated soluble groups of finite rank", *Bull. London Math. Soc.* 16 (1984), pp. 43-46.

[10] J.R.J. Groves, "Finitely presented centre-by-metabelian groups", *J. London Math. Soc.* (2) 18 (1978), pp. 65-69, MR58:28187.

[11] O.G. Harlampovich, "A finitely presented solvable group with unsolvable word problem", *Izvestia Akad. Nauk. Ser. Mat.* 45 (1981), no. 4, pp. 852-873 (Russian), MR82m:20036.

[12] G. Higman, "Subgroups of finitely presented groups", *Proc. Royal Soc. London Ser. A* 262 (1961), pp. 455-475, MR24:A152.

[13] G. Higman, *Finitely presented infinite simple groups* (Notes on Pure Math., I.A.S., Australian Nat. Univ., 1974).

[14] G.P. Kukin, "The equality problem and free products of Lie algebras and of associative algebras", *Sib. Math. Jour.* 24 (1983), no. 2, pp. 85-96.

[15] R.C. Lyndon and P.E. Schupp, *Combinatorial Group Theory* (Springer-Verlag, Berlin, 1977), MR58:28182.

[16] R. Strebel, "Subgroups of finitely presented centre-by-metabelian groups", to appear in *Proc. London Math. Soc.*

[17] R.J. Thompson, "Embeddings into finitely generated simple groups which preserve the word problem", *Word Problems II: The Oxford Book* (Adian, Boon and Higman, eds., North-Holland Pub. Co., Amsterdam, 1980), pp. 401-441, MR81k:20050.

University of California, Irvine
Irvine, California 92717
U.S.A.

PROCEEDINGS OF 'GROUPS — KOREA 1983' 20F99, 20E34
KYOUNGJU, August 1983

THE CONCEPT OF "LARGENESS" IN GROUP THEORY II

M. Edjvet and Stephen J. Pride

1. Introduction

In [20] the second author defined the concept of a large property of groups in order to give precision to vague notions of "largeness" and "smallness" in infinite group theory. The point of view adopted in [20] has proved quite fruitful, and in this paper we will survey most of the results concerning the concept of "largeness" in group theory which have been obtained so far. Some new results will also be presented here. In addition, several open problems will be raised and discussed.

A large property was defined in [20] to be a group-theoretic property $\underline{\underline{P}}$ satisfying: (i) $\underline{\underline{P}}$ is closed under taking pre-images; (ii) if H is of finite index in G, then G has $\underline{\underline{P}}$ if and only if H has $\underline{\underline{P}}$; (iii) the trivial group 1 does not have $\underline{\underline{P}}$. Thus (i) says that if a group is "large" in the sense of having $\underline{\underline{P}}$, then every group which can be mapped homomorphically onto $\underline{\underline{P}}$ is also "large", while (ii) says that two groups which "differ by a finite bit at the top" are either both "large" in the sense of having $\underline{\underline{P}}$ or are both "not large". It is not hard to show that (ii) is equivalent to: (ii)' if H is a normal subgroup of finite index in G, then G has $\underline{\underline{P}}$ if and only if H has $\underline{\underline{P}}$.

Since writing [20], and inspired by some remarks in the doctoral dissertation of Atkinson [1], the second author has reflected that it is rather unnatural to consider only "finite differences at the top"; one should also allow "finite differences at the bottom". This has led us to introduce a further postulate into the definition of a large property, namely the dual of (ii)' (via the duality between subgroups and factor groups): (iv) if G has a finite normal subgroup N, then G has $\underline{\underline{P}}$ if and only if G/N has $\underline{\underline{P}}$. (Note that in the presence of (i) the "if" part of (iv) can be dropped.) For technical reasons it is also convenient to dispense with (iii). If 1 has $\underline{\underline{P}}$ then $\underline{\underline{P}}$ is simply the property of being a group, and so $\underline{\underline{P}}$ can be regarded as the trivial large property in this case. If 1 does not have $\underline{\underline{P}}$ then no finite group can have $\underline{\underline{P}}$ by (ii).

In this survey, whenever we quote a result or refer to a proof from a previous paper concerning "largeness", it will mean that the proof carries over with only minor modifications to conform to our new definition.

Many well-known group-theoretic properties are large properties (see §2 for some examples); a good "typical" property to keep in mind is SQ-universality. Given a group A we can define the large property $\underline{\underline{L}}(A)$ generated by A (see §2). This notion enables us to compare two groups. We write $A \preccurlyeq G$ (" G is larger than A ") if G has $\underline{\underline{L}}(A)$, $G \simeq A$ (" G is equally as large as A ") if $A \preccurlyeq G$ and $G \preccurlyeq A$, $A \prec G$ (" G is strictly larger than A") if $A \preccurlyeq G$ but $A \not\simeq G$. This gives a partial ordering — the largeness ordering, also denoted $\preccurlyeq$ — on the collection L of $\simeq$-equivalence classes $[A]$. We let F denote the set of those $[A]$ with A finitely generated, and L_κ (κ an infinite cardinal) the set of those $[A]$ with $|A| \leqslant \kappa$. Then F and L_κ have largest elements $[F_2]$, $[F_\kappa]$ respectively, where for a cardinal c , F_c denotes the free group of rank c .

Given a group G we have the principal ideal $\mathrm{Id}(G)$ generated by $[G]$. If G is finitely generated then we also have the principal filter $\mathrm{Fil}_{\mathsf{F}}(G)$ generated by $[G]$ in F . If $|G| \leqslant \kappa$ we have the principal filter $\mathrm{Fil}_\kappa(G)$ generated by $[G]$ in L_κ . (see §3 for details.)

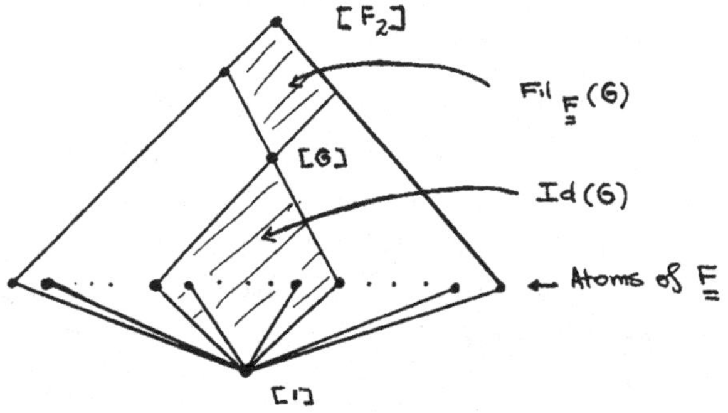

One can get an idea of the "size" of a group by looking at the principal ideal and principal filters it generates. For example, a group G should be regarded as being "small" if $\mathrm{Id}(G)$ is finite, or at least if $\mathrm{Id}(G)$ has finite height. If $\mathrm{Id}(G)$ satisfies one (or both) chain conditions then G should be considered as being "not too large". As regards filters, we conjecture that if $G \prec F_2$ (resp. $G \prec F_\kappa$) then $\mathrm{Fil}_{\mathsf{F}}(G)$ (resp. $\mathrm{Fil}_\kappa(G)$) does not satisfy the maximum chain condition (see Problem 3 below).

In §3 we discuss properties of $(\mathsf{L}, \preccurlyeq)$, and obtain some information concerning principal ideals and filters. If G is finitely generated (resp. $|G| \leqslant \kappa$) we show how the structure of $\mathrm{Id}(G)$ is related to the number of atoms lying *inside* $\mathrm{Id}(G)$, while the structure of $\mathrm{Fil}_{\mathsf{F}}(G)$ (resp. $\mathrm{Fil}_\kappa(G)$) is related to the number of atoms of F (resp. L_κ) lying *outside* $\mathrm{Id}(G)$ (see Theorems 2 and 3). We deduce that $\mathrm{Fil}_{\mathsf{F}}(G)$ does not satisfy the maximum chain condition if G is not SQ-universal (see Theorem 4). We also give a second criterion for $\mathrm{Fil}_F(G)$ not to satisfy the maximum chain condition (Lemma 2); we show that this criterion is satisfied for certain two-generator one-relator groups, for example, the groups $\langle a, t;\ t^{-1}a^{l}t = a^{m}\rangle$ where l, m are relatively prime. Several other results concerning $(\mathbb{L}, \preccurlyeq)$ are given in §3.

In §4 we look at various groups G which are traditionally thought of as being "small", and obtain information concerning $\mathrm{Id}(G)$. Among the groups we consider are polycyclic-by-finite groups (Theorem 5).

In §5 we discuss the maximum and minimum chain conditions for certain principal ideals.

Finally in §6 we summarize the results which have been obtained concerning the determination of those finitely generated groups which are "as large as they can possibly be", that is, those groups which lie in the class $[F_2]$.

Acknowledgements

The authors are grateful to Drs K. Brown and J. Howie for many interesting discussions and useful remarks. In particular, they thank K. Brown for his comments concerning Example 4.3. The first author acknowledges the financial assistance of a Glasgow University Postgraduate Award. Some of the new results contained in this paper will appear in the first author's doctoral dissertation.

Notation

$\mathbb{Z}$, $\mathbb{Z}^+$, $\mathbb{Z}_2$, $\mathbb{Q}$, $\mathbb{C}$ will denote the integers, positive integers, cyclic group or order 2, rationals (thought of as a group under addition), complex numbers, respectively. The highest common factor of two non-zero integers m, n will be denoted by $\mathrm{hcf}(m, n)$.

We write $A \leqslant_f B$, $A \trianglelefteq_f B$ to indicate that A is respectively a subgroup, normal subgroup of finite index in B. We write $A \trianglelefteq^2 B$ if there is a group C with $A \trianglelefteq C \trianglelefteq B$.

Let G be a group and let $x, y \in G$. The conjugate $y^{-1}xy$ of x by y will be denoted by x^y, and the commutator $x^{-1}y^{-1}xy$ of x and y will be denoted by $[x, y]$; $\zeta(G)$ will denote the centre of G, and G^{ab} will denote the abelianization of G (that is, the factor group of G by its derived group $[G, G]$). If $H \leqslant G$ then $\mathrm{Cor}(H)$ will denote the intersection of all the conjugates of H in G. Thus $\mathrm{Cor}(H) \trianglelefteq G$. If $H \leqslant_f G$ then $\mathrm{Cor}(H) \trianglelefteq_f G$, and so we have that *every subgroup of finite index in* G *contains a subgroup of finite index which is normal in* G. This result will be used often, nearly always without comment.

If $\underline{Q}$ is a group-theoretic property then $\neg\,\underline{Q}$ will denote the negation of $\underline{Q}$.

A group is called just-infinite if it is infinite, but all its proper quotients are finite. The class consisting of those groups with every proper subnormal subgroup of finite index will be denoted by $\underline{\underline{D}}_2$.

The restricted wreath product of A and B will be written $A \wr B$.

If c is a cardinal, F_c will denote the free group of rank c .

When using the Reidemeister-Schreier method (in Example 3.2) the notation of [16, §2.3] will be employed without further comment.

2. Large properties and the "largeness ordering"

A group-theoretic property $\underline{\underline{P}}$ will be called a large property if it satisfies:

(LP1) If G has $\underline{\underline{P}}$ and K maps onto G then K has $\underline{\underline{P}}$.

(LP2) If $H \leqslant_f G$ then G has $\underline{\underline{P}}$ if and only if H has $\underline{\underline{P}}$.

(LP3) If N is a finite normal subgroup of a group G with $\underline{\underline{P}}$ then G/N has $\underline{\underline{P}}$.

Using the fact that a subgroup of finite index in a group A contains a subgroup of finite index which is normal in A , we easily see that (LP2) is equivalent to:

(LP2)' If $H \trianglelefteq_f G$ then G has $\underline{\underline{P}}$ if and only if H has $\underline{\underline{P}}$.

Note also that in the presence of (LP1), (LP3) is equivalent to:

(LP3)' If N is a finite normal subgroup of G , then G has $\underline{\underline{P}}$ if and only if G/N has $\underline{\underline{P}}$.

The conditions (LP2)', (LP3)' are dual to each other. They say that if two groups differ by finite "bits" at the top or bottom, then the groups are either both "large" in the sense of having $\underline{\underline{P}}$, or are both "not large".

A useful method for obtaining examples of large properties is given by the following observation. If $\underline{\underline{Q}}$ is a group-theoretic property closed under taking homomorphic images, normal subgroups of finite index, extensions by finite groups, extending finite groups by $\underline{\underline{Q}}$-groups, then $\neg\, \underline{\underline{Q}}$ is a large property.

We now give some examples of large properties.

EXAMPLE 2.1. *Having cardinality greater than or equal to* κ *, where* κ *is an infinite cardinal.*

EXAMPLE 2.2. *The property of not being generated by less than* κ *elements, where* κ *is an infinite cardinal.*

EXAMPLE 2.3. SQ(κ)*-universality, where* κ *is an infinite cardinal.*

A group A is said to be SQ(κ)-universal if every group of cardinality less than or equal to κ is embeddable in a quotient of A . For a verification of

(LP2) see [18]. To verify (LP3), suppose N is a finite normal subgroup of G and assume G is SQ(κ)-universal. Let C be a group with $|C| \leqslant \kappa$. Embed C in a simple group S of cardinality κ [11]. By assumption there is a quotient $\bar{G}$ of G containing a copy S_0 of S. If $\bar{N}$ denotes the image of N then $\bar{N} \cap S_0 = 1$, so S_0 naturally injects into the quotient $\bar{G}/\bar{N}$ of G/N.

EXAMPLE 2.4. *The property of having a free subgroup of rank* κ, *where* κ *is some cardinal.*

EXAMPLE 2.5. *Satisfying no non-trivial law.*

EXAMPLE 2.6. *Non-amenability.*

A group G is said to be amenable if there is a finitely additive translation-invariant measure μ on the set of all subsets of G such that $\mu(G) = 1$. See [7] for further details.

EXAMPLE 2.7. *Having exponential growth.*

We extend the definition [17] to infinitely generated groups. A group G will be said to have exponential growth if there is a finite subset Γ of elements of G such that the following holds: if $\gamma_\Gamma(r)$ $(r \geqslant 0)$ denotes the number of distinct elements of G which can be expressed as words of length less than or equal to r in the elements of Γ then there are constants α, p with $\alpha > 0$, $p > 1$ such that $\gamma_\Gamma(r) \geqslant \alpha p^r$.

See [20, p.305] for a verification of (LP2). To verify (LP3) assume G has exponential growth, so that, using the above notation we have $\gamma_\Gamma(r) \geqslant \alpha p^r$. Now $N \cap \operatorname{sgp}\Gamma$ is finite of order n say. Let ϕ be the natural map of G onto G/N. Then $\gamma_{\Gamma\phi} \geqslant \alpha p^r/n$ and so G/N has exponential growth.

EXAMPLE 2.8. $\neg$ max, $\neg$ min, $\neg$ max-n, $\neg$ min-n, $\neg$ max-sn, $\neg$ min-sn.

See [24, Lemma 1.48, Corollary] and [25, pp.66-68].

EXAMPLE 2.9. *Let* $\underline{\underline{Q}}$ *be a group-theoretic property which is closed under taking homomorphic images and extending finite groups by* $\underline{\underline{Q}}$*-groups; then the property of not being locally* $\underline{\underline{Q}}$*-by-finite is a large property.*

For a verification of (LP2) see [20, p.305]. To see (LP3) suppose that K is a finite normal subgroup of the group G and G/K is locally $\underline{\underline{Q}}$-by-finite. If we let L be a finitely generated subgroup of G then LK/K is $\underline{\underline{Q}}$-by-finite and so L has a normal subgroup H of finite index such that $H/L\cap K$ has the property $\underline{\underline{Q}}$, which means H does since $\underline{\underline{Q}}$ is closed under extending finite groups by $\underline{\underline{Q}}$-groups.

Given a group A we can define the large property $\underline{\underline{L}}(A)$ generated by A as follows. A group G has $\underline{\underline{L}}(A)$ if and only if there is a finite sequence of groups, $A = G_0, G_1, G_2, \ldots, G_{n-1}, G_n = G$ where for $1 \leq i \leq n$:

(i) G_{i-1} is a homomorphic image of G_i,

or (ii) $G_{i-1} \leqslant_f G_i$,

or (iii) $G_i \leqslant_f G_{i-1}$,

or (iv) G_{i-1} has a finite normal subgroup N and $G_{i-1}/N \cong G_i$.

The following Lemma is very useful, and will be used time and time again in the sequel, *nearly always without comment.*

LEMMA 1. (a) *G has $\underline{\underline{L}}(A)$ if and only if there are groups B , C , K with K finite, $K \trianglelefteq B \leqslant_f A$ and $C \leqslant_f G$ such that B/K is a homomorphic image of C .*

(b) *It may be assumed in* (a) *that either C is normal in G , or B and K are normal in A .*

Proof. (a) It suffices to show that if two groups G_{i-1} , G_i are related in one of the ways (i)-(iv), and if G_{i-1} satisfies the conclusion of (a), then so does G_i (the required result then following by induction). This is easily proved.

(b) If C is not normal in G then replace C by $\mathrm{Cor}(C)$ and replace B by B_1 , where B_1/K is the image of $\mathrm{Cor}(C)$ under the homomorphism of C onto B/K .

If B is not normal in A replace B by $\mathrm{Cor}(B)$ and K by $K \cap \mathrm{Cor}(B)$. Finally, suppose $B \trianglelefteq_f A$. Then K has finitely many conjugates $K_1, \ldots, K_m$, say in A , each of which is a finite normal subgroup of B . Then $\mathrm{sgp}\{K_1, \ldots, K_m\}$ is a finite normal subgroup of A contained in B .

This completes the proof.

If G has $L(A)$ we write $G \succcurlyeq A$ (or $A \preccurlyeq G$), and we say that G is *larger than* A . If $G \succcurlyeq A$ and $A \succcurlyeq G$ then we say that G and A are *equally large*, written $G \simeq A$. If $G \succcurlyeq A$ but $G \not\simeq A$ we say that G is *strictly larger than* A , written $G \succ A$ (or $A \prec G$). The relation $\preccurlyeq$ is a quasi-order on the class of groups, and $\simeq$ is the induced equivalence relation. We denote by $[G]$ the $\simeq$-equivalence class containing G . There is then an induced partial order (also denoted $\preccurlyeq$) on the $\simeq$-equivalence classes. This partially ordered structure will be denoted by $(\mathsf{L}, \preccurlyeq)$. Note that $(\mathsf{L}, \preccurlyeq)$ has a smallest element, the $\simeq$-equivalence class consisting of all finite groups.

Perhaps it is worth pointing out that if $\underline{\underline{P}}$ is a large property and A has $\underline{\underline{P}}$, then any group larger than A also has $\underline{\underline{P}}$. Thus two groups which are equally large have the same large properties.

3. Properties of the largeness ordering

We will let F denote the collection of those elements $[G]$ of L with G finitely generated, and for κ an infinite cardinal we let L_κ denote the totality of those $[G]$ where $|G| \leqslant \kappa$.

An obvious question is whether $\preccurlyeq$ is an upper or lower semilattice ordering on

L , or on any of the subsets of L defined above. The answer is "no", even for F .

EXAMPLE 3.1 (L.G. Kovacs). *Let* A , B *be distinct, finite, non-abelian simple groups, and let* S *be an infinite finitely generated simple group. Let* $H_1 = (A \wr S) \times (B \wr S)$, $H_2 = (A \times B) \wr S$, $K_1 = A \wr S$, $K_2 = B \wr S$. *Then* $[H_1]$, $[H_2]$ *are distinct minimal upper bounds for* $[K_1]$, $[K_2]$. *Dually,* $[K_1]$, $[K_2]$ *are distinct maximal lower bounds for* $[H_1]$, $[H_2]$.

Another example, involving polycyclic groups (due to Atkinson [1]) will be described in §4.

It is well known that any partially ordered set can be embedded in a lattice [6]. It is interesting to speculate whether the new elements which one must adjoin to, say, $(\mathsf{F}, \preccurlyeq)$ to get a lattice can be given any significant interpretation.

For a group G , $\mathrm{Id}(G)$ will denote the principal ideal of $(\mathsf{L}, \preccurlyeq)$ generated by $[G]$. Thus the elements of $\mathrm{Id}(G)$ are those $[H] \preccurlyeq [G]$. Since F_n ($2 \leqslant n$, n finite) has (free) subgroups of finite index of arbitrarily large finite rank, and since every finitely generated group is a homomorphic image of a finitely generated free group, we see that

$$(\mathsf{F}, \preccurlyeq) = \mathrm{Id}(F_2) = \mathrm{Id}(F_3) = \ldots = \mathrm{Id}(F_n) = \ldots .$$

Also, if κ is an infinite cardinal then $(\mathsf{L}_\kappa, \preccurlyeq) = \mathrm{Id}(F_\kappa)$.

If G is finitely generated then we write $\mathrm{Fil}_{\mathsf{F}}(G)$ for the principal filter of $(\mathsf{F}, \preccurlyeq)$ generated by $[G]$. Thus the elements of $\mathrm{Fil}_{\mathsf{F}}(G)$ are those $[H]$ in F with $[H] \succcurlyeq [G]$. If κ is an infinite cardinal and $|G| \leqslant \kappa$ we write $\mathrm{Fil}_\kappa(G)$ for the principal filter of $(\mathsf{L}_\kappa, \preccurlyeq)$ generated by $[G]$.

We define the height of G , $\mathrm{ht}G$, to be the maximum length of a chain in $\mathrm{Id}(G)$, if this maximum exists, and ∞ otherwise. We say that G satisfies max-$\preccurlyeq$ (min-$\preccurlyeq$) if $\mathrm{Id}(G)$ satisfies the maximum (minimum) chain condition. Note that the properties "not having finite height", $\neg$ max-$\preccurlyeq$, $\neg$ min-$\preccurlyeq$, are large properties.

PROBLEM 1. *If* $\mathrm{ht}G < \infty$ *then are all maximal chains in* $\mathrm{Id}(G)$ *of the same length?*

PROBLEM 2. *Is it possible for* $\mathrm{ht}G$ *to be infinite but all chains in* $\mathrm{Id}(G)$ *to be finite?*

The next question is of considerable interest. A finitely generated group is "as large as it can possibly be" if $[G] = [F_2]$. One can now ask whether there are any groups which are "almost as large as they can possibly be", that is, is there a group G such that $G \prec F_2$ but there is no group H with $G \prec H \prec F_2$? Such a group would be highly interesting. We note that any two such groups G_1 , G_2 would have to be equally large, for $G_1 \simeq G_1 \times G_2 \simeq G_2$ by Lemma 3 (ii) below. This makes it seem likely that such groups do not exist, and we conjecture that this is the case.

PROBLEM 3. *Show that whenever* $G \prec F_2$ *there is a group* H *with* $G \prec H \prec F_2$.

An equivalent formulation is:

PROBLEM 3'. *Show that whenever* $G \prec F_2$ *then* $\mathrm{Fil}_F(G)$ *does not satisfy the maximum chain condition.*

It will be shown below (see Theorem 4) that if G is a finitely generated group which is not $SQ(\aleph_0)$-universal, then $\mathrm{Fil}_F(G)$ does not have the maximum chain condition.

Another class of groups where Problem 3' has a positive solution is given by the following result.

LEMMA 2. *Let* G *be a finitely generated group with the following property:*

(†) *there is a positive integer* k *such that for every subgroup* $H \leqslant_f G$ *the torsion free rank of* H^{ab} *is less than* k .

Then we have the properly ascending chain:

$$G \prec G \times \mathbb{Z}^k \prec G \times \mathbb{Z}^{2k} \prec \ldots \prec G \times \mathbb{Z}^{nk} \prec \ldots \prec F_2 .$$

This is not hard to prove (using Lemma 1), and is left to the reader.

Lemma 2 appears rather special, but its hypotheses are satisfied by some rather interesting groups. For example the Higman group

$$\langle x, y, z, t;\ x^{-1}yx = y^2,\ y^{-1}zy = z^2,\ z^{-1}tz = t^2,\ t^{-1}xt = x^2\rangle$$

(which is $SQ(\aleph_0)$-universal [27]) has no proper subgroups of finite index [12] so trivially satisfies (†) with $k = 1$.

Here are some examples of one-relator groups satisfying (†) (with $k = 2$).

EXAMPLE 3.2. *Let* $G = \langle a, t;\ t^{-1}a^l t = a^m\rangle$ where l, m are non-zero integers *with* $\mathrm{hcf}(l, m) = 1$. *We show that* (†) *holds with* $k = 2$. *It is enough to show that if* $N \trianglelefteq_f G$ *then* N *has a subgroup* M *of finite index whose abelianisation has torsion-free rank* 1 .

The proof is based on the proof of [20, Theorem 3.4]. Since N is of finite index in G , t and a define elements of finite order, say n , $p \in \mathbb{Z}^+$ respectively, in G/N . Then $\mathrm{hcf}(l, p) = \mathrm{hcf}(m, p) = 1$. For example, suppose $p = n_1 d$, $l = n_2 d$ with $d \geqslant 1$. Since $\mathrm{hcf}(l, m) = 1$ there are integers λ , μ such that $\lambda l + \mu m = 1$. Then

$$a^{n_1} = a^{\lambda n_1 l} a^{\mu n_1 m} = (a^p)^{\lambda n_2} (t^{-1}a^p t)^{\mu n_2} \in N ,$$

so $n_1 = p$ and $d = 1$. Let M be the normal closure of $\{t^n, a^p\}$ in G . Then G/M has presentation

$$\langle a, t;\ t^{-1}a^l t = a^m,\ t^n,\ a^p\rangle$$

and aM generates a normal subgroup of order precisely p in G/M. So a Schreier transversal for M in G is $\{t^i a^j;\ 0 \le i < n,\ 0 \le j < p\}$, and M can be generated by the elements

$$a_i = t^i a^p t^{-i} \qquad (0 \le i < n),$$

$$s(i, j) = t^i a^j t\ \overline{t^i a^j t}^{-1} \qquad (0 \le i < n,\ 0 \le j < p).$$

Denote $s(n-1, 0)$ $(= t^n)$ by y. We have the following relations holding:

$$a_0^l = a_1^m,\ a_1^l = a_2^m,\ \dots,\ a_{n-2}^l = a_{n-1}^m,$$

$$y^{-1} a_0^{l^n} y = a_0^{m^n}.$$

Thus if we abelianise M, a_0 and hence each a_i has finite order. Therefore we need only consider $\hat{M}$, the quotient of M obtained by putting each a_i equal to 1. Now the defining relators of M consist of those $s(i, j)$ for which $t^i a^j t\ \overline{t^i a^j t}^{-1}$ is freely equal to 1, together with the rewrites of the relators

$$t^i a^j t\ a^m t^{-1} a^{-l} a^{-j} t^{-i} \qquad (0 \le i < n,\ 0 \le j < p)$$

of G. Working in $\hat{M}$, these rewrites turn out to be

$$s(i, j) s(i, j+m)^{-1}$$

where $j+m$ is reduced modulo p to lie between 0 and $p-1$. Since $\mathrm{hcf}(m, p) = 1$ it follows that for $i = 0, \dots, n-1$

$$s(i, 0) = s(i, 1) = \dots = s(i, p-1).$$

Now $t^i t\ \overline{t^i t}^{-1}$ is freely equal to 1 unless $i = n-1$. Consequently $\hat{M}$ is infinite cyclic generated by $s(n-1, 0) = y$.

EXAMPLE 3.3 (Edjvet and Howie). *Let* $G_0 = \langle a, t_0;\ t_0^{-1} a^2 t_0 = a^3 \rangle$, *and inductively define* G_{i+1} *to be the* HNN *extension*

$$\langle G_i, t_{i+1};\ t_{i+1}^{-1}\, a\, t_{i+1} = t_i \rangle .$$

(Note that G_{i+1} *is a one-relator group.) Then* G_{i+1} *satisfies* (†) *with* $k = 2$.

For let H be a subgroup of finite index in G_{i+1}. Then H contains the normal closure B_i of the base of G_{i+1}. For let N be a normal subgroup of finite index of G_{i+1} contained in H. It suffices to show that $a \in N$. Suppose aN (and hence $t_0 N$, since a and t_0 are conjugate in G_{i+1}) has order n in G_{n+1}/N. Since $t_0^{-n} a^{2^n} t_0^n = a^{3^n}$, $a^{3^n - 2^n} \equiv 1 \bmod N$. Thus $3^n \equiv 2^n \bmod n$. This implies that $n = 1$. For suppose that $n > 1$, and let p be the smallest prime divisor of n. Then $p \ge 5$. Write $n = p^s q$ with $s > 0$ and q coprime to p.

Since $2^{p^s} \equiv 2 \bmod p$ and $3^{p^s} \equiv 3 \bmod p$, we have $2^q \equiv 3^q \bmod p$. Hence in GF(p), $(2.3^{-1})^q = 1$. But $(2.3^{-1})^{p-1} = 1$, and so q has a prime divisor less than p, a contradiction.

We now see that $H = \mathrm{sgp}\{B_i, t_{i+1}^m\}$ for some $m \geq 1$. Now the derived group of H is B_i, since the derived group contains $a = [a^2, t_0]$ and $a^{t_{i+1}} = t_i$. Thus H^{ab} is infinite cyclic.

A group A is said to be atomic if $[A]$ is an atom in $(L, \preccurlyeq)$ (that is, ht$A = 1$). Atomic groups are called minimal in [20], [21]. Examples of atomic groups are $\underline{\underline{D}}_2$-groups and countable direct powers of finite simple groups. The abelian atomic groups are $D \oplus B$, $\mathbb{Z} \oplus B$, $P \oplus B$, where B is a finite group, D is quasicyclic, and P is a countable p-primary group [21, Theorem 3].

PROBLEM 4. *Describe the finitely generated atomic groups.*

It seems reasonable to suggest that the finitely generated atomic groups are precisely the groups with the structure:

finite

finitely generated infinite $\underline{\underline{D}}_2$

finite

This would be the case if one could show that

(*) every finitely generated just-infinite group satisfies max-sn .

For suppose (*) holds, and let A be a finitely generated atomic group. Then A has a just-infinite quotient B [24, Lemma 6.17]; thus $B \simeq A$. By Example 4.2 below, B has a $\underline{\underline{D}}_2$-subgroup B_0 of finite index. Since $B_0 \simeq A$, Lemma 1 implies that there are groups $N \lhd A_0 \lhd_f A$ with N finite and A_0/N a homomorphic image of a subgroup B_1 of finite index in B_0. Since $B_1 \in \underline{\underline{D}}_2$ [29], $A_0/N \cong B_1$.

PROBLEM 4'. *Does every finitely generated just-infinite group satisfy* max-sn *?*

In [13] Hurley claims to have constructed a countable atomic group which is SQ($\aleph_0$)-universal. Details of this construction have not yet been published.

We will see shortly (Theorem 2) that the structure of a principal ideal is influenced by the number of atoms lying in the ideal.

THEOREM 1. *Let* $[A_1], \ldots, [A_n]$ *be atoms, and let* $G = A_1 \times \ldots \times A_n$.

(i) *If* $H \preccurlyeq G$ *then* $H \simeq A_{i_1} \times \ldots \times A_{i_r}$ *where* $\{i_1, \ldots, i_r\} \subseteq \{1, \ldots, n\}$. *In particular* ht$G \leq n$.

(ii) *If the* $[A_i]$ *are distinct then* Id(G) *is isomorphic to the lattice of all subsets of* $\{1, \ldots, n\}$. *In particular,* ht$G = n$.

Proof. (i) is a restatement of Theorem 4.5 of [20].

(ii) For J a subset of $\{1, \ldots, n\}$ let

$$G_J = \prod_{j \in J} A_j .$$

By (i) the elements of $\mathrm{Id}(G)$ are the $[G_J]$. Moreover, again by (i), if $[G_J] \preccurlyeq [G_K]$ then $[G_J] = [G_{K'}]$ with $K' \subseteq K$. Now, in fact, $K' = J$. For suppose there was an element $j \in J \backslash K'$. Then $[A_j]$ would be an atom lying below $[G_{K'}]$ whereas, by (i), the atoms lying below $[G_{K'}]$ are $[A_k]$, $k \in K'$. We now see that $[G_J] \preccurlyeq [G_K]$ if and only if $J \subseteq K$, and the result follows.

THEOREM 2. *If* G *is a group and* $[A_1], \ldots, [A_n]$ are distinct atoms lying in $\mathrm{Id}(G)$ *then* $G \succcurlyeq A_1 \times \ldots \times A_n$. *In particular* $\mathrm{ht}G \geqslant n$.

For a proof see [21, Theorem 8].

An obvious consequence of Theorem 1 (ii) and Theorem 2 is that if a principal ideal $\mathrm{Id}(G)$ contains infinitely many atoms then G does not satisfy max-$\preccurlyeq$. Another consequence is that if a group G has finite height then $\mathrm{Id}(G)$ contains at most $\mathrm{ht}G$ atoms. One can ask whether the entire ideal generated by a group of finite height must be finite. An example due to Hales [10] shows that this is not the case. Hales gives an example of a torsion-free abelian group of height 4 with infinitely many distinct (up to $\simeq$) groups of height 3 lying below it.

PROBLEM 5. *If* G *is a* *<u>finitely generated</u>* *group of finite height then is* $\mathrm{Id}(G)$ *finite?*

We have seen that information about a principal ideal can be obtained by looking at atoms *in* the ideal. We now show, dually, that we can obtain information about principal filters by looking at atoms lying *outside* principal ideals. The results we present have some bearing on Problem 3.

LEMMA 3. *Let* F *be a non-abelian free group.*

(i) *If* $G = A \times B$ *and* $\phi : G \to F$ *is an epimorphism then* $\mathrm{Ker}\phi$ *contains one of* A, B.

(ii) *If* $A_1, \ldots, A_n \prec F$ *then* $A_1 \times \ldots \times A_n \prec F$.

Proof. (i) Suppose, by way of contradiction, that $A\phi$, $B\phi$ were both non-trivial. Let $x_1, x_2 \in A\phi \backslash \{1\}$, $y_1, y_2 \in B\phi \backslash \{1\}$. Then $[x_1, y_1] = [x_2, y_2] = 1$, so there are elements $u, v \in F$ and integers k, l, m, n such that $x_1 = u^k$, $y_1 = u^l$, $x_2 = v^m$, $y_2 = v^n$. Now $[u^k, v^n] = 1$, so $[u, v] = 1$ by Exercise 4, p.4: of [16]. Thus $[x_1, x_2] = [y_1, y_2] = 1$. Since x_1, x_2, y_1, y_2 were arbitrary, this means that F is abelian, a contradiction.

(ii) The result follows by induction once it has been shown for $n = 2$. Suppose $A_1 \times A_2 \simeq F$. Passing to subgroups of finite index in A_1, A_2, F, if necessary, we may assume that $A_1 \times A_2$ maps onto F. Thus by (i) one of A_1, A_2 maps onto F.

LEMMA 4. *Let* $[A_1]$, $[A_2]$ *be distinct atoms. If* $G \not\succcurlyeq A_2$ *then* $G \times A_1 \not\succcurlyeq A_2$.

Proof. Suppose, by way of contradiction, that $G \times A_1 \succcurlyeq A_2$. Passing to subgroups of finite index in G, A_1 and replacing A_2 by a group equally as large as it if necessary, we may suppose that $G \times A_1$ maps onto A_2. Let N be the kernel of this homomorphism, and let N_1 be the projection of N onto A_1. If $|A_1 : N_1| < \infty$ then GN has finite index in $G \times A_1$, and so

$$\frac{G}{G\cap N} \cong \frac{GN}{N} \leqslant_f \frac{G\times A_1}{N} \cong A_2 .$$

Thus $G \succcurlyeq A_2$, a contradiction. Suppose then that $|A_1 : N_1| = \infty$. Then

$$A_2 \cong \frac{G\times A_1}{N} \succcurlyeq \frac{G\times A_1}{GN_1} \cong \frac{A_1}{N_1} \simeq A_1 ,$$

so $A_2 \simeq A_1$, again a contradiction.

THEOREM 3. *Let* $[G]$ *lie in* F *(resp.* L_κ, κ *an infinite cardinal), and suppose* $[A_1], [A_2], \ldots, [A_n]$ *are distinct atoms in* F *(resp.* L_κ*) lying outside* $\mathrm{Id}(G)$. *Let* $[F]$ *denote the maximum element of* F *(resp.* L_κ*), where* F *is a free group of appropriate rank. Then*

(i) $[G] \prec [G \times A_1] \prec [G \times A_1 \times A_2] \prec \ldots \prec [G \times A_1 \times \ldots \times A_n] \prec [F]$,

(ii) *there is no unrefinable chain*

$$[G] = [G_0] \prec [G_1] \prec \ldots \prec [G_n] = [F] .$$

Proof. (i) follows from Lemmas 3, 4 using induction.

(ii) Suppose such a chain existed. For $1 \leqslant i \leqslant n-1$ let

$$J_i = \{j : G_i \not\succcurlyeq A_j\} .$$

Then $J_1 \supseteq J_2 \supseteq \ldots \supseteq J_{n-1}$. Now $|J_1| = n$ and, by (i), $|J_{n-1}| \leqslant 1$. Thus there is an i such that $|J_{i-1} \backslash J_i| \geqslant 2$. We now prove the following, which contradicts that our chain is unrefinable: if $K \prec H$ and there are distinct atomic groups M_1, M_2 with $M_1, M_2 \not\preccurlyeq K$ and $M_1, M_2 \preccurlyeq H$ then there is a group L with $K \prec L \prec M$ Passing to a subgroup of finite index in H if necessary, we have $H/N \simeq M_1$, $H/D \simeq K$ for certain normal subgroups N, D. If $|H : ND| = \infty$ then $M_1 \simeq H/ND \preccurlyeq H/D \simeq K$, a contradiction. Thus $|H : ND| < \infty$, when we have

$$K \simeq \frac{ND}{D} \preccurlyeq \frac{ND}{N\cap D} \preccurlyeq H .$$

Now the first inequality is strict, otherwise we would have

$$K \simeq \frac{ND}{N\cap D} \not\geqslant \frac{ND}{N} \simeq M_1 \quad .$$

Also, the second inequality is strict, for otherwise we would have

$$M_2 \leqslant H \simeq \frac{ND}{N\cap D} \simeq \frac{ND}{N} \times \frac{ND}{D} \simeq M_1 \times K$$

which contradicts Lemma 3.

THEOREM 4. *If* G *lies in* F *and* $\mathrm{Fil}_{\mathrm{F}}(G)$ *satisfies the maximum chain condition then* G *is* $\mathrm{SQ}(\aleph_0)$*-universal.*

Note that the Higman group discussed earlier shows that the converse of the above theorem does not hold.

It is well known (see [15], p.190 for example) that any countable group A can be embedded in a finitely generated simple group. We will need to use the fact that A can be embedded into infinitely many non-isomorphic finitely generated simple groups. In fact A can be embedded into uncountably many such groups. For there are uncountably many non-isomorphic non-abelian finitely generated simple groups S_i $(i \in I)$ not embeddable into A. If $i \neq j$ then $A \times S_i$ is not isomorphic to $A \times S_j$. For if $\theta : A \times S_i \to A \times S_j$ were an isomorphism then either $S_i\theta \cap S_j = 1$ in which case $S_i\theta \trianglelefteq A$ (since S_j is non-abelian simple), a contradiction, or $S_j \leqslant S_i\theta$ in which case $S_j = S_i\theta$, again a contradiction. Now embed the $A \times S_i$ into finitely generated simple groups; obviously uncountably many such groups will be required.

Proof of Theorem 4. Let A be a countable group. By Theorem 3 there are only finitely many atoms in F lying outside $\mathrm{Id}(G)$. Since infinite simple groups are atomic, it follows from our previous remark that there is an infinite simple group S with $[S] \leqslant [G]$ and A embeddable into S.

Now, by Lemma 1, there exist subgroups $L \trianglelefteq H \trianglelefteq_f G$ with $H/L \cong S$. Let $g_1, \ldots, g_n$ be a transversal for $N_G(L)$ in G and put $L_i = g_i^{-1} L g_i$ $(i = 1, \ldots, n)$. Since $H \trianglelefteq G$ each L_i is normal in H and $H/L_i \cong S$. Put

$$N = \bigcap_{i=1}^{n} L_i$$

so that $N \trianglelefteq G$. We will show that H/N contains a copy of S.

Let J be minimal with respect to $J \subseteq \{1, \ldots, n\}$ and

$$\bigcap_{j\in J} L_j = N .$$

If $|J| = 1$ the result is obvious. Suppose $|J| = m > 1$ and assume without loss of generality that $J = \{1, \ldots, m\}$. Then $N = L_1 \cap M$ where

$$M = \bigcap_{i=2}^{m} L_i .$$

Now L_1M/L_1 is a normal subgroup of H/L_1 and, if trivial we get $L_1 \cap M = M$, contradicting the minimality of J . Therefore, since H/L_1 is simple, we have $L_1M = H$ so that $H/N = L_1M/L_1 \cap M \cong H/L_1 \times H/M$, which completes the proof.

Remark. The analogue of Theorem 4 for non-finitely generated groups would be that if $[G]$ lies in L_κ and $\mathrm{Fil}_\kappa(G)$ satisfies the maximum chain condition then G is $\mathrm{SQ}(\kappa)$-universal. One could prove this in a similar way to the proof of Theorem 4 provided one knew that any group of cardinality less than or equal to κ could be embedded into infinitely many non-isomorphic infinite simple groups of cardinality less than or equal to κ . This is certainly true when $\kappa = \aleph_0$ and is presumably true in general.

Recall that if $(\underline{\underline{X}}, \leqslant)$, $(\underline{\underline{Y}}, \leqslant)$ are two partially ordered sets, then their direct product $(\underline{\underline{X}}, \leqslant) \times (\underline{\underline{Y}}, \leqslant)$ consists of all pairs (x, y) , $x \in \underline{\underline{X}}$, $y \in \underline{\underline{Y}}$ with order defined by $(x, y) \leqslant (x', y')$ if and only if $x \leqslant x'$, $y \leqslant y'$.

If G , H are two groups then the existence of an isomorphism $\mathrm{Id}(G \times H) \cong \mathrm{Id}(G) \times \mathrm{Id}(H)$ is in some sense a reflection of a certain "independence" between G and H . As a trivial example, we see from Theorem 1 that any two distinct (up to $\simeq$) atomic groups are "independent" in this sense. Here is another fairly straightforward result.

LEMMA 5. *If G , H are torsion groups such that the order of every element of G is coprime to the order of every element of H , then*

$$\mathrm{Id}(G \times H) \cong \mathrm{Id}(G) \times \mathrm{Id}(H) .$$

Proof. Any subgroup of finite index in $G \times H$ contains a subgroup $G_1 \times H_1$ with $G_1 \leqslant_f G$, $H_1 \leqslant_f H$. Moreover, a homomorphic image of $G_1 \times H_1$ is (isomorphic to) the direct product of a homomorphic image of G_1 and a homormorphic image of H_1 . We thus see that each element of $\mathrm{Id}(G \times H)$ has a representative from the set

$$\underline{\underline{C}} = \{A \times B : A \text{ (resp. } B \text{) is a homomorphic image of a subgroup of finite index in } G \text{ (resp. } H \text{)}\} .$$

Suppose $A \times B$, $C \times D \in \underline{\underline{C}}$ with $A \times B \precsim C \times D$. By Lemma 1 there is a finite normal subgroup N of $A \times B$ such that some subgroup of finite index in $(A \times B)/N$ is a homomorphic image of some subgroup of finite index in $C \times D$. Let N_A , N_B be the projections of N onto A , B respectively. Then $(A \times B)/N$ maps homomorphically onto $A^* \times B^*$ where $A^* = A/N_A$, $B^* = B/N_B$. Now a subgroup of finite index in $C \times D$ contains $C_1 \times D_1$ for some $C_1 \leqslant_f C$, $D_1 \leqslant_f D$. Some homomorphic image of $C_1 \times D_1$ is of finite index in $A^* \times B^*$. The image $\bar{C}_1$ of C_1 must lie in A^* , and the image $\bar{D}_1$ of D_1 must lie in B^*. Moreover $|A^* : \bar{C}_1| < \infty$, $|B^* : \bar{D}_1| < \infty$. Thus $A \lessdot C$, $B \lessdot D$.

It follows from the previous paragraph that the mapping of $\mathrm{Id}(G \times H)$ to $\mathrm{Id}(G) \times \mathrm{Id}(H)$ given by

$$[A \times B] \mapsto ([A], [B]), \quad A \times B \in \underline{\underline{C}}$$

is well-defined and order-preserving. It is clearly bijective with order-preserving inverse.

4. Some groups of finite height

It is natural to expect that groups which are traditionally thought of as being "small" should have finite height.

We describe some well known groups of finite height.

EXAMPLE 4.1. *Free abelian groups of finite rank.*

Let $A(n)$ $(n \geqslant 0)$ denote the free abelian group of rank n. It is easily shown that if $[H] \lessdot [A(n)]$ then $[H] = [A(m)]$ for some $m \leqslant n$; also if $r < s$ then $[A(r)] \lessdot [A(s)]$. Thus $\mathrm{Id}(A(n))$ is a chain of length n:

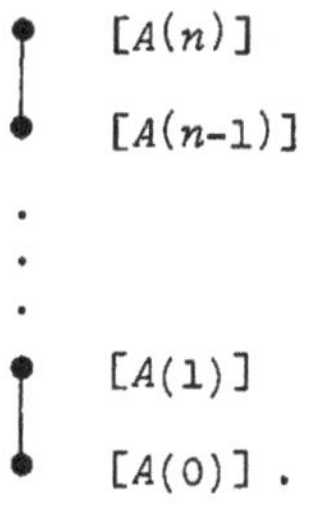

EXAMPLE 4.2. *Just-infinite groups satisfying* max-sn.

Let G be such a group. We give the structure of $\mathrm{Id}(G)$. Suppose first that the Baer radical (i.e. the subgroup generated by the cyclic subnormal subgroups) of G is trivial. In [31], Wilson defines a congruence $\sim$ on the lattice of subnormal subgroups of G as follows: for K, L subnormal in G, $K \sim L$ if and only if $K \cap L$ has finite index in both K and L. The quotient of the lattice of subnormal subgroups by $\sim$ is called the *structure lattice* of G. Now, Wilson shows that G has a subnormal $\underline{\underline{D}}_2$-subgroup $M \trianglelefteq^2 G$ such that the normal closure of M in G is the direct product $M_1 \times \dots \times M_n$ of the distinct conjugates of M in G. If K is subnormal in G then $K \sim M_{i_1} \times \dots \times M_{i_r}$ for some subset $\{i_1, \dots, i_r\}$ of $\{1, \dots, n\}$. Hence the structure lattice of G is isomorphic to the lattice of subsets of $\{1, \dots, n\}$. Now it turns out that $\mathrm{Id}(G)$ is a chain of length n:

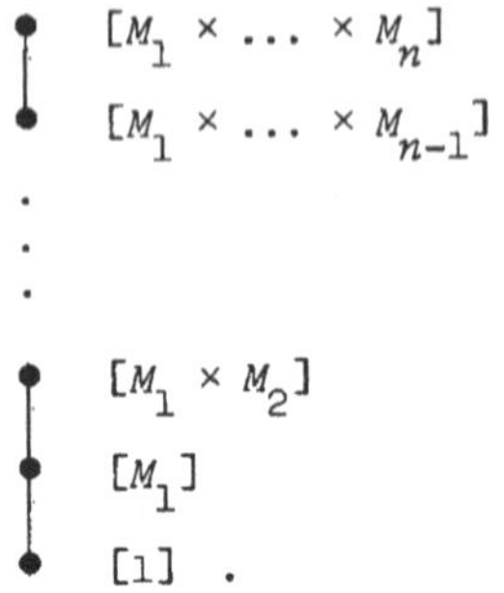

To see this first note that if $[D] \prec [G]$ then D is equally as large as some subnormal subgroup of G (and so D is equally as large as $M_1 \times \ldots \times M_r$ for some $0 \leqslant r \leqslant n$, by what was said above). For there is a homomorphism from a normal subgroup G_0 of finite index in G onto a group equally as large as D . Let N be the kernel of this homomorphism. Then N is subnormal in G and so has a near-complement C (i.e. C is subnormal, $\mathrm{sgp}\{N, C\} = N \times C$, $|G : N \times C| < \infty$) [31, Theorem 3B]. Clearly $G_0/N \simeq C \cap G_0$.

It remains to show that if $H = M_1 \times \ldots \times M_p$, $L = M_1 \times \ldots \times M_q$, $1 \leqslant p < q \leqslant n$ then $H \prec L$. Suppose not. Noting that G (and therefore L) has no non-trivial finite subnormal subgroups [31, Proposition 3], it follows that there must be a homomorphism from a subgroup of finite index in H onto a subgroup of finite index in L . It is shown in [20, p.326] that this is impossible.

If G has non-trivial Baer radical then [31, Theorem 2] G is a finite extension of a free abelian group of finite rank, n say. We now appeal to Example 4.1.

EXAMPLE 4.3. *Polycyclic-by-finite groups.*

Let G be a polycyclic-by-finite group. A well known number associated with G is its Hirsch number $\mathrm{h}G$. A less well known number which can be associated with G is its *plinth length* ρG , which will be defined shortly. (These two numbers coincide if and only if G is nilpotent-by-finite; in general $\rho G \leqslant \mathrm{h}G$). We will show:

THEOREM 5. *If* G *is a polycyclic-by-finite group then*

(i) $\mathrm{ht}G = \rho G$;

(ii) *all maximal chains in* $\mathrm{Id}(G)$ *have the same length.*

We will also describe an example due to Atkinson [1] which shows that *the principal ideal generated by a polycyclic group need not be a lattice.*

A group A is a plinth in G if [26, p.396]: A is a non-trivial torsion-free abelian normal subgroup of G ; $A \otimes_{\mathbb{Z}} \mathbb{Q}$ is irreducible as a $\mathbb{Q}H$-module under conjugation for every subgroup H of finite index in G .

By Lemma 1.4, p.532 of [19] every infinite polycyclic-by-finite group has a subgroup of finite index with a plinth. Suppose $H_1 \leqslant_f G$ and A_1 is a plinth in H_1. Then either H_1/A_1 is finite, in which case we go no further, or there exists an H_2 of finite index in H_1 and a plinth A_2/A_1 in H_2/A_1. Continuing in this way we obtain a series

$$1 = A_0 < A_1 < \ldots < A_n \leqslant \ldots \leqslant H_n \leqslant_f \ldots \leqslant_f H_2 \leqslant_f H_1 \leqslant_f G ,$$

where A_i/A_{i-1} is a plinth in H_i/A_{i-1} $(1 \leqslant i < n)$, and H_n/A_n is finite. The number n is the same for all such series, and is called the plinth length of G, denoted ρG. We note some properties of ρG (in this connection see [26, pp.396-397]).

(a) $\rho G = 0$ if and only if G is finite;

(b) if $H \cong G/N$ then $\rho H \leqslant \rho G$, with equality if and only if N is finite;

(c) if H is of finite index in G then $\rho H = \rho G$;

(d) if H is of finite index in G and A is a plinth in H, then $\rho H/A = \rho G - 1$.

We will use these facts without further comment. We will also make use of the fact (again without further comment) that if $H \preccurlyeq G$ then there exist $H_1 \leqslant_f H$, $G_1 \leqslant_f G$ with H_1, G_1 torsion-free and $H_1 \cong G_1/N$ (this follows from Lemma 1 and the well known fact that polycyclic groups have torsion free subgroups of finite index). If N is non-trivial (i.e. if N is infinite) then $H \prec G$. We conclude in particular that two polycyclic groups are equally large if and only if they are commensurable (that is, have isomorphic subgroups of finite index).

Proof of Theorem 5. (i) The proof is by induction on $\mathrm{h}G$. If $\mathrm{h}G = 0$ then G is finite, and the result holds in this case. Assume $\mathrm{h}G > 0$, and suppose there is a chain $[G_m] \prec \ldots \prec [G_1] \prec [G]$. We can assume without loss of generality that there is a homomorphism of G onto G_1 with infinite kernel. Then $\mathrm{h}G_1 < \mathrm{h}G$ and $\rho G_1 < \rho G$. By induction, $m-1 \leqslant \mathrm{ht}G_1 = \rho G_1$, and so $\mathrm{ht}G \leqslant \rho G$. To obtain the reverse inequality, let $H \leqslant_f G$ and A a plinth in H. Then $\mathrm{h}H/A < \mathrm{h}H = \mathrm{h}G$ and so $\rho G = \rho H/A + 1 = \mathrm{ht}H/A + 1$ by induction. Since A is infinite $\mathrm{ht}H/A < \mathrm{ht}G$, thus $\rho G \leqslant \mathrm{ht}G$.

(ii) It suffices to show that if $H \prec G$ and there is no L with $H \prec L \prec G$ then $\mathrm{ht}H = \mathrm{ht}G - 1$. For (ii) then follows by an easy induction argument on $\mathrm{ht}G$. We can assume that H is a homomorphic image of G: $H \cong G/N$ with N torsion-free. By considering a derived series for N, there exists $A_1 \leqslant N$ with A_1 a non-trivial torsion-free abelian normal subgroup of G. From the proof of Lemma 1.4, p.532 of [19], there exist $A_0 \leqslant A_1$ and $G_0 \leqslant_f G$ with A_0 a plinth in G_0. We therefore have $G_0/A_0 \succcurlyeq G/A_1 \succcurlyeq G/N$. Since A_0 is infinite, $G_0/A_0 \prec G_0 \simeq G$ and so $G_0/A_0 \simeq H$. By (i) we have $\mathrm{ht}H = \mathrm{ht}G_0/A_0 = \rho G_0/A_0$. But $\rho G_0/A_0 = \rho G_0 - 1 = \rho G - 1 = \mathrm{ht}G - 1$, as required.

Atkinson [1] has given examples of torsion-free polycyclic groups H , K , A , B where: $[A]$, $[B]$ are incomparable; $[H]$, $[K]$ are incomparable $[H], [K] \succcurlyeq [A], [B]$; there is no $[L]$ with $[H], [K] \succ [L] \succ [A], [B]$. Thus if $G = H \times K$ then $\mathrm{Id}(G)$ is not a lattice.

We first give a general construction due to Atkinson. Let A, B be groups with normal subgroups M, N respectively, such that A/M , B/N are both isomorphic to a group C . Let $\theta : C \to A/M$, ϕ $C \to B/N$ be specific isomorphisms. Choose generating sets $Y' = \{y'_\alpha\}$, $Z' = \{z'_\alpha\}$ of M, N respectively, and let presentations on these generating sets be

$$M = \langle Y; S\rangle , \quad N = \langle Z; T\rangle ,$$

where $Y = \{y_\alpha\}$, $Z = \{z_\alpha\}$ are in obvious 1:1 correspondence with Y', Z'. Choose a presentation $\langle X; R\rangle$ for C . For each $x \in X$ choose representatives a, b for the corresponding cosets of M in A , N in B under the isomorphisms θ, ϕ. Now

$y'^{a_\beta}_\alpha = w_{\alpha\beta}(y'_\gamma)$ an element of M written as a word in its generators, for $y'_\alpha \in Y'$, a_β corresponding to $x_\beta \in X$;

$z'^{b_\beta}_\alpha = v_{\alpha\beta}(z'_\gamma)$ an element of N written as a word in its generators, for $z'_\alpha \in Z'$, b_β corresponding to $x_\beta \in X$;

$r_\alpha(a_\beta) = u_\alpha(y'_\gamma)$ in M , $r_\alpha(x_\beta) \in R$;

$r_\alpha(b_\beta) = u'_\alpha(z'_\gamma)$ in N , $r_\alpha(x_\beta) \in R$.

Then

$$A = \langle Y, X; S, y_\alpha^{x_\beta} = w_{\alpha\beta}(y_\gamma), r_\alpha(x_\beta) = u_\alpha(y_\gamma) \text{ for all } y_\alpha \in Y, x_\beta \in X, r_\alpha \in R\rangle ,$$

$$B = \langle Z, X; T, z_\alpha^{x_\beta} = v_{\alpha\beta}(z_\gamma), r_\alpha(x_\beta) = u'_\alpha(z_\gamma) \text{ for all } z_\alpha \in Z, x_\beta \in X, r_\alpha \in R\rangle .$$

We define $A_\theta \bigvee_\phi B$ by

$$A_\theta \bigvee{}_\phi B = \langle Y, Z, X; S, T, [y_\alpha, z_\beta], y_\alpha^{x_\beta} = w_{\alpha\beta}, z_\alpha^{x_\beta} = v_{\alpha\beta}, r_\alpha = u_\alpha u'_\alpha , \text{ for all } y_\alpha \in Y, z_\alpha, z_\beta \in Z, x_\beta \in X, r_\alpha \in R\rangle .$$

Note that $\mathrm{sgp}Y$, $\mathrm{sgp}Z$ are normal in this group, and there are obvious epimorphisms of $A_\theta \bigvee_\phi B$ onto A, B with kernels $\mathrm{sgp}Z$, $\mathrm{sgp}Y$ respectively. In particular, $[A_\theta \bigvee_\phi B]$ is an upper bound for $[A]$, $[B]$.

We now give Atkinson's example. (However, details of proof are omitted. For further information see [1].)

Let A be the split extension of the free abelian groups of rank 2 with basis w, x by the free abelian group of rank 2 with basis y, z where each of y, z acts via the automorphism $w \mapsto wx$, $x \mapsto w^2 x$. Thus

$$A = \langle w, x, y, z;\ [w, x], [y, z], y^{-1}wy = wx, z^{-1}wz = wx , \\ y^{-1}xy = w^2x, z^{-1}xz = w^2x\rangle .$$

Let B be the split extension of the free abelian group with basis a, b by the free abelian group with basis c, d where c acts via $a \mapsto ab$, $b \mapsto b$ and d acts via the identity. Thus

$$B = \langle a, b, c, d;\ [a, b], [c, d], c^{-1}ac = ab, c^{-1}bc = b , \\ d^{-1}ad = a, d^{-1}bd = b\rangle .$$

Let C be the free abelian group of rank 2 with basis r, s . We have the three isomorphisms:

$$\theta : C \to \frac{A}{\mathrm{sgp}\{w, x\}} , \quad r \mapsto y\ \mathrm{sgp}\{w, x\}, \quad s \mapsto z\ \mathrm{sgp}\{w, x\},$$

$$\phi : C \to \frac{B}{\mathrm{sgp}\{a, b\}} , \quad r \mapsto c\ \mathrm{sgp}\{a, b\}, \quad s \mapsto d\ \mathrm{sgp}\{a, b\},$$

$$\psi : C \quad \frac{B}{\mathrm{sgp}\{b, d\}} , \quad r \mapsto a\ \mathrm{sgp}\{b, d\}, \quad s \mapsto c\ \mathrm{sgp}\{b, d\},$$

Let $H = A_\theta \bigvee {}_\phi B$, $K = A_\theta \bigvee {}_\psi B$. Then $[H]$, $[K]$ are both upper bounds for $[A]$, $[B]$.

To show that $H \neq K$ it must be shown that the groups are not commensurable. Atkinson shows that the centre of a subgroup of finite index in H has Hirsch number 1 , whereas the centre of a subgroup of finite index in K has Hirsch number at least 2 . Also, the centre of a subgroup of finite index in A has Hirsch number 1, whereas the centre of a subgroup of finite index in B has Hirsch number at least 2, so $A \neq B$. Finally, it must be shown that there is no L with $H, K \succ L \succ A, B$. Suppose such an L existed. The first thing to do is to find a "nicer" group L^* with $H, K \succ L \succcurlyeq L^* \succ A, B$. To do this, note that L, A, B have subgroups L_0, A_0 , B_0 of finite index such that $A_0 \cong L_0/M_0$, $B_0 \cong L_0/N_0$ for certain normal subgroups M_0 , N_0 . Put $L^* = L_0/M_0 \cap N_0$. If $M = M_0/M_0 \cap N_0$, $N = N_0/M_0 \cap N_0$ then (i) $MN \cong M \times N$, and (ii) M, N are isomorphic to normal subgroups M_1, N_1 of B_0, A_0 respectively, and $A_0/N_1 \cong B_0/M_1 \cong L^*/MN$. Atkinson shows that $|\zeta(A_0/N_1)| < \infty$ whereas $|\zeta(B_0/M_1)| = \infty$, a contradiction.

EXAMPLE 4.4. *Divisible abelian groups.*

A divisible abelian group G is (isomorphic to) the direct sum of quasicyclic groups and copies of $\mathbb{Q}$.

Suppose that G is the direct sum of finitely many, say m , quasicyclic groups; assume there are m_1 of type $\mathbb{Z}p_1^\infty$, m_2 of type $\mathbb{Z}p_2^\infty$, ..., m_r of type $\mathbb{Z}p_r^\infty$ where the p_i are distinct primes, the m_i are greater than 0 , and $m_1 + m_2 + \dots + m_r = m$. Then $\mathrm{Id}(G)$ *is isomorphic to the direct product of chains of lengths* $m_1, m_2, \dots, m_r$; *in particular*, $\mathrm{ht}G = m$, *and* $\mathrm{Id}(G)$ *is a lattice.* To prove this it suffices, by Lemma 5, to deal with the case $r = 1$. Write $G = \Sigma_1^m D_i$ where each D_i is a copy of $\mathbb{Z}_{p_1^\infty}$. If $A \preccurlyeq G$ then $A \simeq \Sigma_1^n D_i$ for some $n \leqslant m$ (since some homomorphic image of G must be equally as large as A). Now it is clear that

$$1 \prec D_1 \prec D_1 \oplus D_2 \prec \ldots \prec D_1 \oplus D_2 \oplus \ldots \oplus D_m ,$$

so the result follows.

Now suppose that G is not the direct sum of finitely many quasicyclic groups. Then G , or some homomorphic image of G , is the direct sum of infinitely many quasicyclic groups. It then follows from the previous paragraph that $\mathrm{Id}(G)$ does not satisfy max-$\preccurlyeq$.

EXAMPLE 4.5. *Some other abelian groups.*

Theorems 5 and 6 of [21] give necessary and sufficient conditions for direct sums of cyclic groups, and for abelian torsion groups, to have finite height.

5. Max-$\preccurlyeq$ and Min-$\preccurlyeq$

Recall that a group G is said to satisfy max-$\preccurlyeq$ (resp. min-$\preccurlyeq$) if there are no properly ascending (resp. descending) chains in $\mathrm{Id}(G)$.

In [21] it is shown that poly-$\underline{\underline{D}}_2$ groups satisfy min-$\preccurlyeq$. Let G be an infinite poly-$\underline{\underline{D}}_2$ group and consider a $\underline{\underline{D}}_2$-series $\underline{\underline{S}}$ for G :

$$\underline{\underline{S}} : G = G_0 \trianglerighteq G_1 \trianglerighteq \ldots \trianglerighteq G_n = 1$$

(i.e. each of the factors G_j/G_{j+1} , $0 \leqslant j < n$, is a $\underline{\underline{D}}_2$-group). Any refinement of this series will have the same number of infinite factors as $\underline{\underline{S}}$. Thus, by Schreier's Theorem, any two $\underline{\underline{D}}_2$-series for G have the same number of infinite factors. We denote this number by $\rho_i(G)$. Let G_k/G_{k+1} be the first infinite factor in $\underline{\underline{S}}$, and let $\rho_f(\underline{\underline{S}})$ be the sum of the orders of the finite factors in the series

$$G_k \trianglerighteq G_{k+1} \trianglerighteq \ldots \trianglerighteq G_n = 1 .$$

We let $\rho_f(G) = \min\{\rho_f(\underline{\underline{S}})\ ;\ \underline{\underline{S}}$ is a $\underline{\underline{D}}_2$-series for $G\}$. We define $\rho(G)$ to be the ordered pair $(\rho_i(G), \rho_f(G))$. If G is finite we define $\rho(G)$ to be $(0, 0)$. Ordered pairs of integers are given the lexicographical ordering.

The following facts are fairly readily established (see [21]). Let G be a poly-$\underline{\underline{D}}_2$ group.

(a) If B is a subgroup of finite index in G , then B is a poly-$\underline{\underline{D}}_2$ group and $\rho_i(B) = \rho_i(G)$, $\rho_f(B) \leqslant \rho_f(G)$ (if $B \trianglelefteq G$ then, in fact, $\rho_f(B) = \rho_f(G)$).

(b) If C is a homomorphic image of G , then C is a poly-$\underline{\underline{D}}_2$ group and $\rho(C) \leqslant \rho(G)$. Moreover if $\rho(C) = \rho(G)$ then C and G are commensurable (and, in particular, equally large).

It follows easily from (a) and (b) that if $H \prec G$ then H is equally as large as a poly-$\underline{\underline{D}}_2$ group H_1 with $\rho(H_1) < \rho(G)$. From this one readily deduces that the ideal generated by a poly-$\underline{\underline{D}}_2$ group satisfies the minimum chain condition. However, it does not follow, as claimed in [21], that such an ideal has finite height. (This was pointed out to the second author by Atkinson.)

PROBLEM 6. *Do* poly-$\underline{\underline{D}}_2$ *groups have finite height?*

PROBLEM 7. *Describe the structure of* poly-$\underline{\underline{D}}_2$ *groups.*

Free products other than $\mathbb{Z}_2 * \mathbb{Z}_2$ are usually thought of as being "quite large". Nevertheless, such groups do not always have $\underline{\underline{L}}(F_2)$. In fact, it is shown in [20] that *a free product of two groups has* $\underline{\underline{L}}(F_2)$ *if and only if either: one of the factors does, or: the factors have proper subgroups of finite index, not both of index two.*

PROBLEM 8. *If* $G = A * B$ *with* A, B *nontrivial and not both of order* 2 *then is it true that* G *does not have* max-$\preccurlyeq$, min-$\preccurlyeq$ *?*

Atkinson [1] has investigated this question with various conditions on A and B. In particular she shows that *if* A, B *are infinite simple groups then* G *does not have* max-$\preccurlyeq$. Indeed, if F is the cartesian subgroup of G (F is free on $\{[a, b] : a \in A\backslash\{1\},\ b \in B\backslash\{1\}\}$), and $S_n(F)$ is the nth term in the derived series of F then we have the properly ascending chain:

$$A \times B = \frac{A * B}{S_1(F)} \prec \frac{A * B}{S_2(F)} \prec \dots \prec \frac{A * B}{S_n(F)} \prec \dots \preccurlyeq A * B .$$

Atkinson obtains two corollaries of the above result.

(a) *If* $G = A * B$ *with* A, B *non-trivial finitely generated groups, not both of order* 2 , *then* G *does not have* max-$\preccurlyeq$.

For let A_1, B_1 be maximal normal subgroups of A, B respectively and let $\bar{A} = A/A_1$, $\bar{B} = B/B_1$. Then $\bar{A}, \bar{B}$ are simple and $G \succcurlyeq \bar{A} * \bar{B}$ so that there is nothing more to prove if $\bar{A}, \bar{B}$ are both infinite. If $\bar{A}$ is finite and $\bar{B}$ is infinite then the normal closure N of $\bar{B}$ in $\bar{A} * \bar{B}$ is the free product of $|\bar{A}|$ copies of B, and since $N \simeq \bar{A} * \bar{B}$ there is nothing more to prove. If $\bar{A}, \bar{B}$ are both finite and not both of order 2 then $\bar{A} * \bar{B}$ has $\underline{\underline{L}}(F_2)$. If $|\bar{A}| = |\bar{B}| = 2$ then one of A_1, B_1, say A_1, is non-trivial. The normal closure of A_1 and B in G is of index 2 and is isomorphic to $A_1 * B * B$. This latter group maps onto $S * (\mathbb{Z}_2 \times \mathbb{Z}_2)$ where S is a non-trivial simple quotient of A_1. Now $S * (\mathbb{Z}_2 \times \mathbb{Z}_2)$ has a subgroup of index 4 isomorphic to $S * S * S * S$ and this group does not satisfy max-$\preccurlyeq$ (if S is finite the group in fact has $\underline{\underline{L}}(F_2)$).

(b) max-$\preccurlyeq$ *and "having finite height" are not recursively decidable properties of finitely presented groups.*

For, $\mathbb{Z}$ has finite height. Also, if A is any non-trivial finitely presented group then $\mathbb{Z} * A$ does not have max-$\preccurlyeq$. Thus max-$\preccurlyeq$ and "having finite height" are incompatible with free products, and so the result follows from [15, p.193].

Atkinson has also shown (under our former definition) that *the free product* $A * B$ *of two infinite simple groups* A, B *does not satisfy* min-$\preccurlyeq$. With a bit of modification, her proof carries over to our new definition, as follows. In [30], Vaughan-Lee gives an infinite sequence of laws

$$\underline{\underline{v}}_1 \subseteq \underline{\underline{v}}_2 \subseteq \dots \underline{\underline{v}}_{i-1} \subseteq \underline{\underline{v}}_i \subseteq \dots$$

such that for each $i > 1$, some element of $\underline{\underline{v}}_i$ is not a consequence of $\underline{\underline{v}}_{i-1}$. Moreover, each of the $\underline{\underline{v}}_i$ defines a variety, all of whose elements are (nilpotent of class 2)-by-(nilpotent of class 2). Let N_i be the derived group of $\underline{\underline{v}}_i(F)$ (where, as before, F is the cartesian subgroup of $A * B$). Since $\underline{\underline{v}}_i(F) \subsetneq \underline{\underline{v}}_{i+1}(F)$, $N_i \subsetneq N_{i+1}$ for each i [2]. Now

$$\frac{A * B}{N_{i+1}} \preccurlyeq \frac{A * B}{N_i}.$$

Suppose these groups were in fact equally large. Since the groups neither have proper subgroups of finite index, nor have proper finite normal subgroups (since F/N_i is torsion free [14, p.116]), it follows from Lemma 1 that there would be an epimorphism from $(A * B)/N_{i+1}$ onto $(A * B)/N_i$. As in [1], one can show that this is impossible.

Analogues of (a) and (b) with max-$\preccurlyeq$ replaced by min-$\preccurlyeq$ follow from the previous paragraph.

One can ask whether the properties max-$\preccurlyeq$, min-$\preccurlyeq$ and also "having finite height" are extension closed. The answer is "no" on all three accounts. Let $\pi = \{p_1, p_2, p_3, \dots\}$ be the set of primes. Let S be a finite non-abelian simple group, and let $A_i = S_{i1} \times S_{i2} \times \dots \times S_{ip_i}$ where $S_{ij} \cong S$. Now let

$$A = \prod_{i=1}^{\infty}{}^{D} A_i$$

and let σ_i be a p_i-cycle in the symmetric group on $\{1, \dots, p_i\}$. Define G to be the extension of A by the infinite cyclic group $\langle t \rangle$, where t acts on A_i by permuting the factors according to the permutation σ_i. Then G *is a split extension of one atomic group by another, but* G *does not satisfy* max-$\preccurlyeq$ *or* min-$\preccurlyeq$. For details see [22]. (An example in [20] claiming to show that max-$\preccurlyeq$ and "having finite height" are not extension closed is incorrect; the group in question actually has height 3.)

6. Groups which are "as large as they can be"

A finitely generated group is "as large as it can be" if it is equally as large as F_2. This is equivalent to the condition that the group has a subgroup of finite index which can be mapped onto F_2 . A group which is as large as F_2 has all the large properties enjoyed by F_2 ; in particular, such a group is $SQ(\aleph_0)$-universal.

It is not unreasonable to expect that a group which has a finite presentation where the number of defining relators is small in comparison with the number of generators should be "large". This point of view has been very rewarding in the investigation of groups which are as large as F_2 .

Let $G = \langle X_1, \ldots, X_n; R_1, \ldots, R_m\rangle$ where the R_i are non-empty cyclically reduced words and $2 \leqslant m \leqslant n$. It was shown in [3] that if $n - m \geqslant 2$ *then* $G \simeq F_2$. In [4] it was conjectured that *if* $n - m \geqslant 1$ *and one of the relators of* G *is a proper power then* $G \simeq F_2$. Some partial results supporting this conjecture were obtained in [4]. The conjecture was proved independently by Stöhr [28] and M. Gromov [8]. (These authors also give alternative proofs of the result of [3].) One now can ask what happens when $m = n$. Guided by the above, it is natural to consider the situation when at least two of the defining relators are proper powers. This is done in [5]. The results are as follows. Assume $m = n$. Write each relator R_i in the form $Q_i^{p_i}$ where Q_i is not a proper power and $p_i \geqslant 1$. Reordering the relators if necessary, we can suppose that $p_1 \geqslant p_2 \geqslant \ldots \geqslant p_m \geqslant 1$. Assume $p_2 > 1$ (i.e. at least two of the relators are proper powers). Let M be the exponent sum matrix of the presentation (the matrix whose (i, j)-th entry is the exponent sum of X_j in R_i) . Then $G \simeq F_2$ *if either* (i) $\det M \neq 0$ *and* $(p_1, p_2, p_3, \ldots, p_m) \neq (2, 2, 1, \ldots, 1)$, *or* (ii) $\det M = 0$ *and there exist* p_i, p_j $(i \neq j)$ *with* $\mathrm{hcf}(p_i, p_j) > 1$. The conditions on the p_i's in (i), (ii) cannot be removed in general, as the groups $\langle x, y; x^2, y^2\rangle$, $\langle x, y; [x, y]^2, [x, y]^3\rangle$ show.

It follows from the previous paragraph that if G is a one-relator group on at least 3 generators, or a one-relator group with torsion on at least 2 generators, then $G \simeq F_2$.

PROBLEM 9. *Let* $G = \langle a, t; R\rangle$ *be a two-generator, torsion-free, one-relator group. Under what conditions is* $G \simeq F_2$ *?*

We have already seen in Examples 3.2, 3.3 some one-relator groups which are not equally as large as F_2 .

THEOREM 6. *Let* $G = \langle a, t; t^{-n} a^k t^n = a^l\rangle$ *where* k, l, n *are non-zero integers. Then* $G \simeq F_2$ *if and only if either* $|n| > 1$ *or* $\mathrm{hcf}(k, l) > 1$.

Proof. Suppose $|n| = 1$ and $\mathrm{hcf}(k, l) = 1$. Then, by Example 3.2, $G \not\simeq F_2$.

Conversely, if $\mathrm{hcf}(k, l) = d > 1$, then the quotient G/M , where M is the normal closure in G of a^d , is a group equally as large as F_2 and the result holds. Suppose now that $|n| > 1$ and $\mathrm{hcf}(k, l) = 1$. We may take $n > 0$. Let N be the normal closure in G of $\{a, t^n\}$. Then N may be presented as

$$N = \langle x_1, \ldots, x_n, z; z^{-1} x_j^k z = x_j^l, 1 \leqslant j \leqslant n\rangle .$$

Let p be an odd prime such that $p > |k|, |l|$. Then $\mathrm{hcf}(p, k) = \mathrm{hcf}(p, l) = 1$. Let $\bar{N}$ be the quotient of N obtained by factoring by the normal closure in N of $\{x_j^p : 1 \leqslant j \leqslant n\}$. Since p, k, l are pairwise coprime we have

$$\bar{N} = \langle x_1, \ldots, x_n, z;\ x_j^p,\ z^{-1}x_jz = x_j^q,\ 1 \leqslant j \leqslant n\rangle ,$$

where $\mathrm{hcf}(p, q) = 1$. Now factor by the normal closure in $\bar{N}$ of z^{p-1} to obtain

$$\bar{\bar{N}} = \langle x_1, \ldots, x_n, z;\ x_j^p,\ z^{p-1},\ z^{-1}x_jz = x_j^q,\ 1 \leqslant j \leqslant n\rangle .$$

Since $q^{p-1} \equiv 1 \bmod p$ we see that $\bar{\bar{N}}$ is a finite extension of the group $G_0 = \langle x_1, \ldots, x_n;\ x_1^p, \ldots, x_n^p\rangle$. Since $n > 1$ and $p > 2$, $G_0 \simeq F_2$ and the result follows.

This proves Theorem 6.

In [20] necessary and sufficient conditions are given for a non-trivial free product to have $\underline{\underline{L}}(F_2)$ (see §5 in connection with this). This result has been generalised by Atkinson [1] to certain amalgamated products. She proves the following: *Suppose* C *does not have* $\underline{\underline{L}}(\mathbb{Z})$. *Then* $A \ast_C B$ *has* $\underline{\underline{L}}(F_2)$ *if and only if there are subgroups* $A_1 \trianglelefteq_f A$, $B_1 \trianglelefteq_f B$ *with* $A_1 \cap C = B_1 \cap C$ *and either* (i) A_1 *or* B_1 *has a quotient isomorphic to* F_2, *or* (ii) $|A : A_1C| \geqslant 2$, $|B : B_1C| \geqslant 2$, *not both* 2. Atkinson also obtains a similar result for HNN extensions: *If* $G = \langle A, t;\ C^t = C\phi\rangle$ *is an* HNN *extension of* A *with associated subgroups* C, $C\phi$ (ϕ *being an isomorphism), and* C *does not have* $\underline{\underline{L}}(\mathbb{Z})$, *then* G *has* $\underline{\underline{L}}(F_2)$ *if and only if there is a subgroup* $A_1 \trianglelefteq_f A$ *with* $(A_1 \cap C)\phi = A_1 \cap C\phi$ *and* $|A : A_1C| \geqslant 2$.

PROBLEM 10. *Obtain necessary and sufficient conditions for an arbitrary amalgamated product* (*or* HNN *extension*) *to have* $\underline{\underline{L}}(F_2)$.

In [9] the following result is proved. *Let* $K = \mathbb{Q}(\sqrt{d})$ *be an imaginary quadratic number field of discriminant* $d < 0$, *and let* Θ *be an order of* K. *Then* $\mathrm{SL}_2(\Theta) \simeq F_2$. When Θ is the maximal order of K a similar result was obtained by Zimmert [32] for certain values of d. The method of proof in [9] is a modification of the methods of Zimmert, together with some additional arguments. In [9] the following interesting result is also noted. *Let* K *be an algebraic number field and let* Θ *be an order of* K. *Then* $\mathrm{SL}_n(\Theta)$ *is* $\mathrm{SQ}(\aleph_0)$*-universal if and only if* $n = 2$ *and either* $K = \mathbb{Q}$ *or* K *is an imaginary quadratic number field.*

Some interesting results concerning the largeness of finitely generated 3-manifold groups and discrete subgroups of $\mathrm{SL}(2, \mathbb{C})$ are given in [23].

We mention two final problems:

PROBLEM 11. *Investigate the largeness of finitely generated small cancellation groups.* (See [20, pp. 315-317] in connection with this problem.)

PROBLEM 12. *What sorts of groups are equally as large as free groups of infinite rank?*

References

[1] J.A. Atkinson, *Large Properties of Groups*, Ph.D thesis, Queen Mary College, London, 1980.

[2] M. Auslander and R.C. Lyndon, "Commutator subgroups of free groups", *Amer. J. Math.* 77 (1955), pp. 929-931, MR17:709.

[3] B. Baumslag and S.J. Pride, "Groups with two more generators than relators", *J. London Math. Soc.* (2) 17 (1978), pp. 425-426, MR58:11137.

[4] B. Baumslag and S.J. Pride, "Groups with one more generator than relators", *Math. Z.* 167 (1979), pp. 279-281, MR81i:20014.

[5] M. Edjvet, "Groups with balanced presentations", to appear in *Archiv der Math.*

[6] H. Gericke, *Lattice Theory* (George G. Harrap, London, Toronto, Wellington, Sydney, 1966), MR36:2534.

[7] F.P. Greenleaf, *Invariant Means on Topological Groups* (Van Nostrand Mathematical Studies No. 16, Van Nostrand-Reinhold Company, New York, Toronto, London, Melbourne, 1969), MR40:4776.

[8] M. Gromov, "Volume and Bounded Cohomology", *Publ. Math.* IHES 56 (1983), pp. 213-307.

[9] F. Grunewald and J. Schwermer, "Free non-abelian quotients of SL_2 over orders of imaginary quadratic number fields", preprint.

[10] A. Hales, "Groups of height four", *J. Austral. Math. Soc.* (Series A) 29 (1980), pp. 297-300, MR81f:20039.

[11] P. Hall, "On embedding a group into the join of given groups", *J. Austral. Math. Soc.* 17 (1974), pp. 434-495, MR51:13055.

[12] G. Higman, "A finitely generated infinite simple group", *J. London Math. Soc.* 26 (1951), pp. 51-64, MR12:390.

[13] B.M. Hurley, "Small cancellation theory over groups equipped with an integer-valued length function", *Word Problems II: The Oxford Book* (Adian, Boone, and Higman, eds., North-Holland, Amsterdam, New York, Oxford, 1980), pp. 157-214, MR81m:20043. [See particularly, p. 207, p. 212.]

[14] L.G. Kovacs, "The thirty-nine varieties", *Math. Scientist* 4 (1979), pp. 113-128, Zbl.416:20018.

[15] R.C. Lyndon and P.E. Schupp, *Combinatorial Group Theory* (Springer-Verlag, Berlin, Heidelberg, New York, 1977), MR58:28182.

[16] W. Magnus, A. Karrass, and D. Solitar, *Combinatorial Group Theory: Presentations of Groups in Terms of Generators and Relations* (Second revised edition, Dover, New York, 1976), MR54:10423.

[17] J. Milnor, "A note on curvature and fundamental group", *J. Differential Geom.* 2 (1968), pp. 1-7, MR38:636.

[18] P.M. Neumann, "The SQ-universality of some finitely presented groups", *J. Austral. Math. Soc.* 16 (1973), pp. 1-6, MR48:11342.

[19] D.S. Passman, *The Algebraic Structure of Group Rings* (Wiley-Interscience, New York, 1977), MR81d:16001.

[20] S.J. Pride, "The concept of 'largeness' in group theory", *Word Problems II: The Oxford Book* (Adian, Boone, and Higman, eds., North-Holland, Amsterdam, New York, Oxford, 1980), pp. 299-335, MR81c:20038.

[21] S.J. Pride, "On groups of finite height", *J. Austral. Math. Soc.* (Series A) 28 (1979), pp. 87-99, MR80m:20028.

[22] S.J. Pride, "On the maximum and minimum chain conditions for the 'largeness' ordering on the class of groups", to appear.

[23] J.G. Ratcliffe, "Euler characteristics of 3-manifold groups and discrete subgroups of SL(2, $\mathbb{C}$)", preprint.

[24] D.J.S. Robinson, *Finiteness Conditions and Generalized Soluble Groups*, Parts 1 and 2 (Ergebnisse der Mathematik und ihrer Grenzgebiete, Bands 62 and 63, Springer-Verlag, New York, Berlin, 1972), MR38:11314, 48:11315.

[25] D.J.S. Robinson, *A Course in the Theory of Groups* (GTM 80, Springer-Verlag, New York, Heidelberg, Berlin, 1982).

[26] J.E. Roseblade, "Prime ideals in group rings of polycyclic groups", *Proc. London Math. Soc.* (3) 36 (1978), pp. 385-447, MR58:10996a.

[27] P.E. Schupp, "Small cancellation theory over free products with amalgamation", *Math. Ann.* 193 (1971), pp. 255-264, MR45:392.

[28] R. Stöhr, "Groups with one more generator than relators", *Math. Z.* 182 (1983), pp. 45-47, MR81i:20044.

[29] C. Tretkoff, "Some remarks on just-infinite groups", *Commun. Algebra* 4 (1976), pp. 483-489, MR63:8269.

[30] M.R. Vaughan-Lee, "Uncountably many varieties of groups", *Bull. London Math. Soc.* 2 (1970), pp. 280-286, MR43:2054.

[31] J.S. Wilson, "Groups with every proper quotient finite", *Proc. Cambridge Philos. Soc.* 69 (1971), pp. 373-391, MR43:338.

[32] R. Zimmert, "Zur SL_2 der ganzen Zahlen eines imaginär-quadratischen Zahlkorpers", *Inventiones Math.* 19 (1973), pp. 73-81, MR47:6883.

Department of Mathematics,
University of Glasgow,
University Gardens,
Glasgow G12 8QW, Scotland

PROCEEDINGS OF 'GROUPS – KOREA 1983'
KYOUNGJU, August 1983

20F99, 20E22

EXTENDING GROUPS VIA TREE AUTOMORPHISMS

Narain Gupta

This lecture is based on recent joint work with Said Sidki of the University of Brasília. Here I will present a new construction of groups using tree automorphisms and discuss some of its applications.

Let $G_1, G_2, \ldots$ be an infinite sequence of (not necessarily distinct) non-trivial groups $G_i = \{e_i, g_i, h_i, \ldots\}$, where e_i is the identity of G_i. Let $X = \{(x_1, \ldots, x_k) \mid k \geq 0, x_i \in G_i\}$ be the set of all ordered k-tuples. Then $(X; \leq)$ is a partially ordered set with $\emptyset < x_1 < (x_1, x_2) < \ldots$; and elements of X may be viewed as the vertices of the infinite descending tree $T(\emptyset)$:

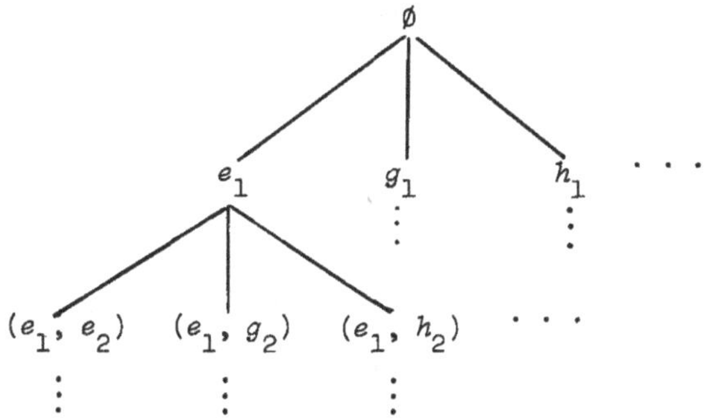

where the vertices at level $k \geq 0$ are the elements of X of length k. If u, v are vertices of length $k \geq 1$, then the subtrees $T(u)$ and $T(v)$, with roots u and v respectively, are isomorphic. Let $A = \mathrm{Aut}(T(\emptyset))$ be the automorphism group of $T(\emptyset)$. Any automorphism $(u; \alpha)$ of $T(u)$, $u \in X$, can be extended to an automorphism $(u; \alpha)^*$ of $T(\emptyset)$ by simply letting $(u; \alpha)^*$ fix every vertex of $T(\emptyset)$ which is outside the subtree $T(u)$. For any $h \in G_k$, $u = (x_1, \ldots, x_{k-1})$, $k \geq 1$, let $(u; h)$ be the automorphism of $T(u)$ defined by

$$(u, x_k, x_{k+1}, \ldots, x_t) \to (u, x_k h, x_{k+1}, \ldots, x_t)$$

for all $t \geq k$ and all $x_i \in G_i$, $i = k, \ldots, t$. Then $(u; h)^* \in \mathrm{Aut}(T(\emptyset))$ and, for any u of length $k-1$, the map $h \to (u, h)^*$ embeds G_k in A. If

$u = (e_1, \ldots, e_{k-1})$, then we shall write h^* for $(u; h)^*$ without ambiguity.

For each $i = 1, 2, \ldots$, let $G_i^{\#} = G_i \backslash \{e_i\}$ and define a subset S of X as:

$$S = S_1 \cup S_2 \cup S_3 \cup \ldots ,$$

where $S_k = (e_1, \ldots, e_{k-1}, G_k^{\#}) = \{(e_1, \ldots, e_{k-1}, x_k) \mid x_k \in G_k^{\#}\}$. Let

$$\alpha : S \to \sum_{i=1}^{\infty} G_i$$

be any function such that $\alpha(S_k) \subseteq G_k$ for all k . Then we may regard α as a formal sum

$$\alpha = \sum_{k=1}^{\infty} \alpha_k ,$$

where $\alpha_k = \alpha\big|_{S_k}$ is the restriction of α on S_k . For every choice of

$$\alpha : S \to \bigcup_{i=1}^{\infty} G_i ,$$

there corresponds an automorphism $\gamma_1 \in A$, given by

$$\gamma_1 = \sum_{k=1}^{\infty} \sum_{s \in S_k} \alpha(s)^* ,$$

where for $s \in S_k$, $\alpha(s)^* = (e_1, \ldots, e_{k-1}; \alpha(s))^* = \gamma_1\big|_S$. If we set

$$\delta_k = \gamma_1\big|_{S_k} = \sum_{s \in S_k} \alpha(s)^* ,$$

then we may write

$$\gamma_1 = \delta_1 + \delta_2 + \ldots ;$$

and

$$\gamma_k = \delta_k + \delta_{k+1} + \ldots = \delta_k + \gamma_{k+1}$$

for all $k = 1, 2, \ldots$. Thus

$$\gamma_k = \sum_{g \in G_k^{\#}} \alpha(e_1, \ldots, e_{k-1}, g)^* + \gamma_{k+1}$$

$$= \sum_{g \in G_k} \alpha(e_1, \ldots, e_{k-1}, g)^*$$

where $\alpha(e_1, \ldots, e_{k-1}, e_n)^* = \gamma_{n+1}$, without ambiguity. With this notation, for any $h \in G_k$, we have

$$h^{*-1}\gamma_k h^* = \sum_{g \in G_k} \alpha(e_1, \ldots, e_{k-1}, gh^{-1})^* .$$

For each $k = 1, 2, \ldots$, the group $E_k = \langle \gamma_k, h^* \mid h \in G_k \rangle$ is called the extension of G_k by γ_k . While the significance of these general extensions is not yet fully understood, the particular case $G = G_1 = G_2 = \ldots$ and

$\alpha(g) = \alpha(e, g) = \alpha(e, e, g) = \dots$, has proven to be of fundamental importance.

Let G be any non-trivial group and let $\alpha : S \to G$ be any function with $\alpha(g) = \alpha(e, g) = \alpha(e, e, g) = \dots$ for all $g \in G^{\#}$. Consider the extension $E = \langle r, G\rangle$, where

$$\gamma = \sum_{k=1} \sum_{s \in S_k} \alpha(s)^* .$$

Then

$$\gamma = \sum_{k=1} \sum_{g \in G^{\#}} \alpha(\underset{\leftarrow\ k-1\ \rightarrow}{e, \dots, e}, g)^*$$

$$= \sum_{g \in G^*} \alpha(g)^* + (e; \gamma)^* ,$$

where $(e; \gamma)^*$ is the automorphism induced by γ on $T(e)$. By setting $(e; \gamma)^* = \alpha(e)^*$, we may conveniently write

$$\gamma = \sum_{g \in G} \alpha(g)^*$$

and for all $h \in G$,

$$\gamma^h = h^{-1}\gamma h = \sum_{g \in G} \alpha(gh^{-1})^* .$$

Let $\Gamma = \langle \gamma^h \mid h \in G\rangle$. Then $E = \Gamma G$ is the semi-direct product of Γ with G. Assume further that $G = \langle \alpha(g) \mid g \in G^{\#}\rangle$ (e.g. $\alpha(g) = g$). For some general properties of E, I refer to the paper [1]. This construction is particularly suitable for producing examples of finitely generated infinite p-groups. For instance, if we take $G = \langle a \mid a^p = 1\rangle$, p odd, $\alpha(a^i) = a^{(-1)^{i+1}}$, $i \not\equiv o(p)$, then $E = \langle \gamma, a\rangle$ is an infinite p-group generated by elements of order E. In addition, it turns out that every finite p-group is embedded in E and that every proper quotient of E is finite. For the case $p = 2$, we take $g = \langle a \mid a^4 = 1\rangle$ and $\alpha(a) = a$, $\alpha(a^2) = 1$, $\alpha(a^3) = a^3$. Then $E = \langle \gamma, a\rangle$ is an infinite 2-group with generators of order 4 and, every finite 2-group is embedded in E. The details of these properties can be found in [2].

References

[1] Narain Gupta, and Said Sidki, "Extensions of groups by tree automorphisms", *Contemporary Mathematics* (1984), to appear.

[2] Narain Gupta, and Said Sidki, "Some infinite p-groups", *Algebra i Logika* 22 (1983), pp. 584-589.

Department of Mathematics,
University of Manitoba,
Winnipeg, Canada

PROCEEDINGS OF 'GROUPS — KOREA 1983' 20E06
KYOUNGJU, August 1983

HNN-CONSTRUCTING FINITE GROUPS*

Verena Huber-Dyson

It seems only fitting that on the occasion of this fine conference we take time out to contemplate the power of that fruitful principle for building infinite groups known as HNN-extension, due to Graham Higman, the late Hanna Neumann and, of course, Bernhard, three of the nicest and most creative people in the field. I will give you a very beautiful, elementary, as well as amusing, illustration. The story in short is this: at the beginning there was just one finite group, the group S_3 of the symmetries of the regular triangle, the smallest non-commutative group. It contains three copies of the smallest non-trivial group, any two of which can be used to form an HNN-extension from which, via a centralizing HNN-extension followed by an identification you arrive at a group that embeds the next symmetric group S_4 as a splitting quotient. Repeating this process all finite symmetric groups are obtained, and so, as subgroups, are all finite groups. All this is seen easily by analysing the group S of permutations of the integers generated by a transposition and the successor. I will introduce the necessary notation, separating the concrete from the abstract, exhibit the crucial presentations and describe the construction but leave the proofs that involve no more than calculations with relators to the reader. At the end I shall mention a few remarkable properties of the group S .

Notation

(a) Generalities

$\langle X \rangle$, the free group on the set X of symbols.

$\langle W \rangle$, $(\langle\langle W \rangle\rangle)$, the (normal) subgroup of $\langle X \rangle$ generated by W , for $W \subseteq \langle X \rangle$.

The *presentation* $\langle X | W \rangle$ is defined by the exactness of the sequence

$$1 \longrightarrow \langle\langle W \rangle\rangle \longrightarrow \langle X \rangle \overset{\langle 1 \rangle}{\longrightarrow} \langle X | W \rangle \longrightarrow 1 ,$$

where $\langle 1 \rangle$ is induced by the identity $1_X : X \to X$.

---

* This note is dedicated to the many Korean students that so graciously and patiently attended all the lectures.

$\langle U\rangle_G$, $(\langle\langle G\rangle\rangle^G)$, the (normal) subgroup of a group G generated by $U \subseteq G$. The label 'G' will be omitted when there is no danger of equivocation, in particular when it is redundant, as in $\langle U\rangle_G$ for a concrete set U of transformations. If the set U generates G then any "naming" $\pi : X \twoheadrightarrow U$, (surjection), has, by virtue of the freedom of $\langle X\rangle$, a unique extension $\langle\pi\rangle$ that completes the diagram

$$\begin{array}{ccc} X & \hookrightarrow & \langle X\rangle \\ \pi \downarrow & & \downarrow \langle\pi\rangle \\ U & \hookrightarrow & G \end{array} .$$

We write $G = g\pi\ \langle X|R\rangle$ and say that $\langle X|R\rangle$ *presents* G via π if $\langle\pi\rangle = \langle\langle R\rangle\rangle$. If $\psi : A \overset{\cong}{\to} B$ is an isomorphism of subgroups of G and c does not occur in G we write $G*^c_\psi$ for the HNN-extension of G by ψ with the stable letter c. If $\langle X|R\rangle$ presents G via π then $\langle X, c|R, \{w^cv^{-1}|w\langle\pi\rangle \in A$ and $v\langle\pi\rangle = w\langle\pi\rangle\psi\}\rangle$ presents $G*^c_\psi$. It suffices to let the w's run through any preimage of a set of generators for A, and we are usually content with the self-explanatory hybrid notation $\langle G, c|\bar{c}Ac = A\psi\rangle$. The importance of $G*^c_\psi$ is that it embeds G, while extending the isomorphism of the subgroups to an inner automorphism. Note that $G*^c_\psi$ is entirely characterized by G and the isomorphism ψ, so that c need only be mentioned when we are dealing with presentations. HNN-extensions by identity morphisms, $G*_{1_A}$, are called centralizing. More generally, if $\psi = \dot{g}|A : A \to A^g$ is the restriction of the inner automorphism induced by g, the HNN-extension $G*^c_\psi$ is the split extension $\langle\langle c^{-1}g\rangle\rangle \rtimes G$.

(b) Specifics

We let i range over the set $\mathbb{Z}$ of the integers and let $n > 0$. First we need a few formal definitions, words and sets of words of $\langle a, b\rangle$:

$$a_i = b^{-i}ab^i\ ,\quad b_n = b^{-(n-1)}(ba)^{n-1} \quad \text{and so} \quad b_1 = 1\ ,\quad b_{n+1} = a_{n-1}b_n = b_n{}^b a$$

$$R_1 = \emptyset\ ,\ R_2 = \{a^2\}\ ,\ R_3 = \{a^2, [b, a, a\ \ a, b]\}\ ,\ R_{n+3} = R_{n+2} \cup \{[b^{n+1}, a, a]\}\ ,$$

where we write $[x, y, z]$ for $[[x, y], z]$, and observe that $a^2 = 1$ implies $[b, a, a] = [a, b]^2$ and so the last relators in the sets R_3 and R_{n+3} can be replaced by $[b, a]^3$ and $[b^{n+1}, a]^2$ respectively.

We obtain relations $E_2 = \{a^2=1\}$ and $E_{n+1} = E_n \cup \{a^b = a^{b_n}\}$, set $S^{(n)} = \langle a, b|R_n\rangle$ and $S^{(\omega)} = \langle a, b|R\rangle$, where $R = \bigcup_1^\infty R_n$, for the abstract presentations, and introduce now some concrete groups of permutations of the integers. If τ denotes the transposition (01), ρ the successor $i \mapsto i+1$ and ζ_n the n-cycle $(0, 1, \ldots, n-1)$ we set $\tau_i = (i, i+1)$ and note that $\tau_i = \rho^{-i}\tau\rho^i$ and $\zeta_n = \tau_{n-1} \cdots \tau_1\tau_0$. For the finite symmetric group of degree n it is a familiar exercise to show that $S_n = \langle\tau, \zeta_n\rangle = \langle\{\tau_i | i < n-1\}\rangle$. These two modes of generating the finite symmetric groups lead to two limits that are to be discussed here, the locally finite group S_∞ of permutations of finite support, and its extension $S = \langle\tau, \rho\rangle$ by an infinite cycle that shifts the τ_i, $\tau_i \mapsto \tau_{i+1}$, by conjugation. It is easy to see that

$$S_\infty = \langle\langle\tau\rangle\rangle^S = \langle\{\tau_i | i \in \mathbb{Z}\}\rangle = \{x \in S \mid \exp(x) < \infty\} ,$$

that the derived group $S' = S'_\infty = \langle\langle[\rho, \tau]\rangle\rangle = \{[x, y] | x, y \in S\}$ is the unique minimal normal subgroup of S as well as of S_∞ and that an element of S belongs to S_∞ if and only if it is conjugate to its inverse. Note that $S_\omega = \bigcup_\omega S_n = \langle\tau_n | n \in \omega\rangle \lneq S_\infty \simeq S_\omega$.

In order to relate the abstract to the concrete we use the mappings $\pi_n : a \mapsto \tau$, $b \mapsto \zeta_n$, and $\beta_n : S_{n-1} \mapsto S_{n-1}^{\zeta n}$, the restriction of conjugation by the n-cycle to the symmetric group of degree $n-1$, and observe that this morphism coincides with the restriction of ρ to S_{n-1}. We define $\pi = \pi_\omega$ and $\beta = \beta_\omega$ by the respective actions of $a \mapsto \tau$, $b \mapsto \rho$, and $\{\tau_n \mapsto \tau_{n+1} | n \in \omega\}$, while we denote by γ the outer automorphism of S_∞ induced by conjugation with ρ .

Now that the tools are assembled, the facts can be stated and the verification left as an easy exercise.

LEMMA. (i) $S_n = g\pi_n\langle a, b\ R_n, b^{-1}b_n\rangle$ *, for all* $n > 1$ *, and* $S = g\pi\langle a, b | R\rangle$.

(ii) $S_n *^b_{\beta_n} \simeq S_\infty^{(n)}$ *by* $\tau \mapsto a$, $\zeta_n \mapsto b_n$ *and* $b \mapsto b$ *, for all* $n > 1$.

(iii) $S_\omega *^b_\beta \simeq S_\infty *^b_\gamma \simeq S$ *, moreover* $S = S_\infty \rtimes \langle\rho\rangle = \langle\langle\rho\rangle\rangle \rtimes \langle\tau\rangle$.

(iv) *S is the direct limit of the chain of epimorphisms* $\varepsilon_n : S^{(n)} \twoheadrightarrow S^{(n+1)}$ *induced by the relation* $a_{n-1}^{-1}aa_{n-1} = a$ *, from* $n = 3$ *on ,* $S^{(n)} = \langle\langle b^{-1}b_n\rangle\rangle \rtimes S_n$ *, and S_ω is the direct limit of the chain of injections* $\iota_n : S_n \to S_{n+1}$.

Observing that each ε_n is the composition of a centralizing HNN-extension and an identification of the stable letter with an element of the base group, one obtains a uniform recipe for constructing all finite symmetric groups, starting with S_3 and using only HNN-extensions and identifications. If you insist you can even start with S_2 . Here is the scheme, for $n \geq 2$, with $\psi_n : a \to a^{b_n}$,

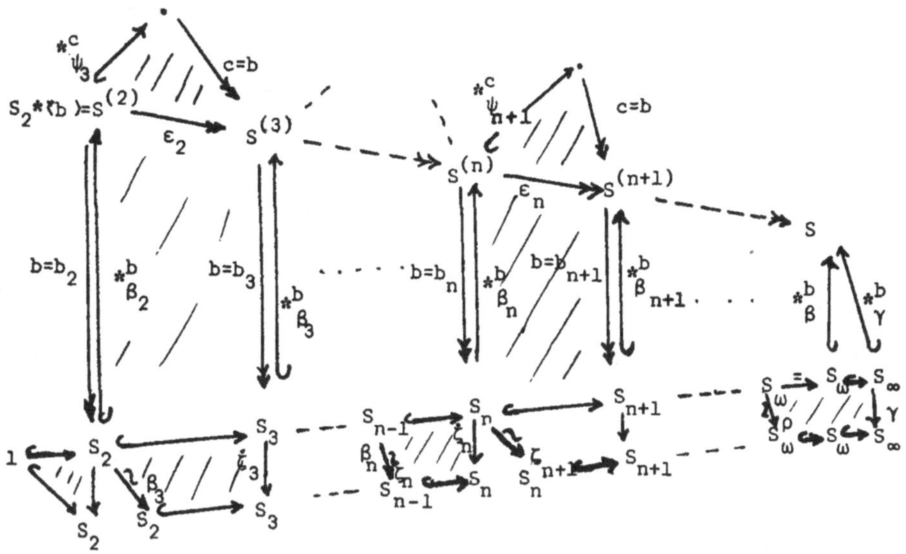

The lemma has a remarkable corollary. By a result of Cohen's [1] all the groups $S^{(n)}$ are residually finite, but then by "general nonsense" the direct limit S is imbeddable in a model of the theory of finite groups, and this in turn means that the elementary theory of the group S contains the universal theory of finite groups. Since S also embeds every finite group we have arrived at the

COROLLARY. *The universal theory of finite groups coincides with the universal theory of the group S of permutations of the integers generated by a transposition and the successor.*

In other words, a finite system of equations and inequations is solvable in some finite group if and only if it is so in S. In [5] Slobodskoĭ proves the undecidability of the universal theory of finite groups, and thus S has an unsolvable equation problem. A deeper analysis of the group S shows that its structure is intimately related to elementary arithmetic. By an existential interpretation, [3], it is possible to code diophantine problems into S and to deduce the undecidability of the equation problem involving constants a and b from Matijasevič's proof [4] of the unsolvability of Hilbert's tenth problem. Moreover, the finite symmetric groups are uniformly elementarily describable in S and so the theory of finite symmetric groups can be relatively interpreted in the theory of S. It should be observed that S is by no means residually finite. In fact, its largest residually finite image is S abelianized, thus illustrating the existence of finitely generated infinitely related not residually finite presentations all of whose finitely presented preimages are residually finite. The fact that Slobodskoĭ's proof establishes the recursive inseparability of the universal theory of periodic groups from the set of finitely refutable universal sentences raises the quest for a universal theorem of finite group theory that fails in some periodic group. One would hardly expect a finitely generated infinite Burnside group to be embeddable in a model of the theory of finite groups and therefore a good place to look for such a sentence would be finite presentations with p-th power relators for some large enough prime.

Returning to the group S, one of the most rewarding endeavours is to describe as much of its structure in the language of first order group theory, for instance to characterize its generators τ and ρ by the simplest possible formulas, in the hope of arriving at Slobodskoĭ's result without the use of Minsky machines. More information is contained in [3] and work is in progress.

References

[1] D.E. Cohen, "Residual finiteness and Britton's Lemma", *J. London Math. Soc.* (2) 16 (1977), pp. 232-234, MR57:3249.

[2] G. Higman, B.H. Neumann, and H. Neumann, "Embedding theorems for groups", *J. London Math. Soc.* 24 (1949), pp. 247-254, MR11:322.

[3] V. Huber-Dyson, "Symmetric groups and the Open Sentence Problem", *Patras Logic Symposium* (G. Metakides, ed., North-Holland Publ. Co., 1982), pp. 159-169.

[4] Ju.V. Matijasevič, "Enumerable sets are Diophantine", *Dokl. Akad. Nauk SSSR* 191 pp. 279-282 (Russian), MR41:3390; English transl., *Soviet Math. Dokl.* 11 pp. 354-358.

[5] A.M. Slobodskoĭ, "Undecidability of the universal theory of finite groups", *Algebra i Logika* 20 (1981), pp. 207-230 (Russian), MR83h:03062; English transl. *Algebra and Logic* 20 (1981), no. 2, pp. 139-156 (1982).

Department of Philosophy,
University of Calgary,
Alberta, Canada T2N 1N4

PROCEEDINGS OF 'GROUPS — KOREA 1983' 20F36, 20F05
KYOUNGJU, August 1983

ANALOGUES OF THE BRAID GROUP

D.L. Johnson

Like the Artin groups of Appel and Schupp [3], which stand in the same relation to the braid group [4] as do Coxeter groups (see [8]) to the symmetric group, the analogues considered here are defined by means of a presentation. The aim of this talk is to give a brief survey of the results obtained so far, by myself and my students at Nottingham, on the structure of these groups. I am grateful to Jens Mennicke for pointing out the close connection which exists between certain of our groups and the mapping class groups of closed orientable surfaces (see [5] for background, and [15], [18], [25] for details), a connection that we propose to investigate further in the near future.

1. Background and notation

In an attempt at self-containment, we shall paraphrase the work of Chow [7] (see also [17] and [11], section 29) on the structure of the braid group.

The algebraic braid group B_n is generated by symbols $x_1, x_2, \ldots, x_{n-1}$ and defined by relations

$$\begin{aligned} x_i x_{i+1} x_i &= x_{i+1} x_i x_{i+1}, \qquad 1 \le i \le n-2 \\ x_i x_j &= x_j x_i \qquad 1 \le i < j-1 \le n-2, \end{aligned} \tag{1_n}$$

called *braid relations*, *commutations*, and abbreviated to $x_i \approx x_{i+1}$, $x_i \sim x_j$, respectively. To avoid triviality, it is customary to assume that $n \ge 3$. The elements of this group are called *braids*, and they act in a natural way on $\pi_1(\mathbb{R}^2 \backslash n \text{ points})$, which is just the free group F_n of rank n. Denoting the free generators by $a_1, \ldots, a_n$,

$$\begin{gathered} x_i \text{ fixes each of } a_1, \ldots, a_{i-1}, a_{i+2}, \ldots, a_n \\ \text{and also the product } a_1 a_2 \ldots a_n, \text{ and sends } a_i \text{ to } a_{i+1}. \end{gathered} \tag{2_n}$$

This induces a homomorphism

$$\rho : B_n \to \mathrm{Aut}\, F_n\,, \qquad (3_n)$$

whose image is called the *geometrical braid group*.

To get at the structure of B_n, a toe-hold is provided by the natural map

$$\left.\begin{aligned} \sigma : B_n &\to S_n \\ x_i &\mapsto (i,\ i+1) \end{aligned}\right\}, \qquad (4_n)$$

onto the *symmetric group* $S_n = \mathrm{Sym}\{1, 2, \ldots, n\}$. If $\mathrm{Stab}(1) = S_{n-1} \leq S_n$, define $P_1 = \rho^{-1}(S_{n-1})$, the group of 1-*pure braids*. The intersection of these is just $\mathrm{Ker}\, \sigma$, called the *unpermuted braid group*, and sometimes denoted by U.

Using the Reidemeister-Schreier process ([11], §12), Chow obtained a presentation for P_1 (which has index n in B_n), from which it is clear that

$$P_1 = M_1] B_{n-1}\,, \qquad (5_n)$$

a split extension of the group M_1 of 1-*smooth* braids (isomorphic to F_{n-1}) by B_{n-1} (generated by $x_2, \ldots, x_{n-1}$), with action given by (3_{n-1}).

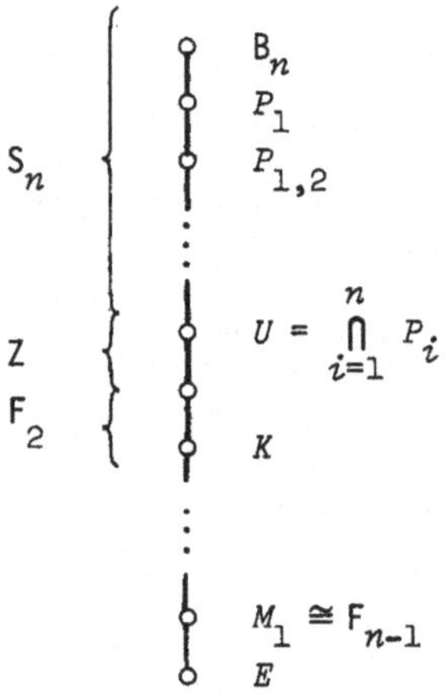

This gives an inductive step in the study of the B_n, and we have the picture indicated on the left. The fact that U is a tower of free groups leads to a *normal form* for braids, whence B_n has soluble word problem. A clever argument with centralisers then shows that the centre of B_n is

$$Z(B_n) = \langle (x_1 \ldots x_{n-1})^n \rangle\,,$$

the infinite cyclic group generated by a "complete twist". Finally, it follows that the homomorphism ρ of (3_n) is one-to-one.

2. Some sidelights

a) B_n has soluble conjugacy problem [9].

b) The upper central sides of B_n stops at $Z(B_n)$, that is $Z(B_n/Z(B_n))$ is trivial, by a slick argument due to H.R. Morton.

c) It turns out that $U/K \cong F_2 \times F_1$, whence B_n is *SQ*-universal (see [22] and [23]), and equally as large as F_2, in the sense of [24].

d) The group $M = \bigcap_{i=1}^{n} M_i$ of *smooth braids* is a subgroup of $M_1 \cong F_{n-1}$ generated modulo any term of the lower central series by those basic commutators that involve *all* the free generators of F_{n-1} ([10], [13], [14]).

e) The action (2_n) induces an action of B_n on the free product $\mathbb{Z}_k^{*n}$ of n copies of the cyclic group of order k. That this action is faithful strengthens the last result of the previous section, and is proved in [6] by topological methods. Using a purely algebraic approach, it is shown in [12] that B_n even acts faithfully on the commutator subgroup of $\mathbb{Z}_k^{*n}$, provided that k is even. Though the method breaks down when k is odd ([19], [20]) the result is true in general, as follows from the Birman-Hilden theorem using an (unpublished) argument of D.J. Collins. The case $k = 2$ leads to a faithful action of B_n on F_{n-1}. Question: Is it true that $n-1$ is the smallest value of m such that B_n embeds an Aut F_m ?

3. Circular braids

Consider the group $\hat{B}_n$ (informally) obtained from B_n by declaring that (in the standard combinatorial picture (see [11] §29 or [16]) the crossing of the 1 and the n strings be independent of the others, which can be guaranteed by interposing an extra 0-ring between it and the others. More precisely, modify the relations (1_{n+1}) by replacing $x_1 \sim x_n$ by $x_1 \approx x_n$. To describe the structure of $\hat{B}_n$, we paraphrase [2].

We expand (5_{n+1}) to the following diagram:

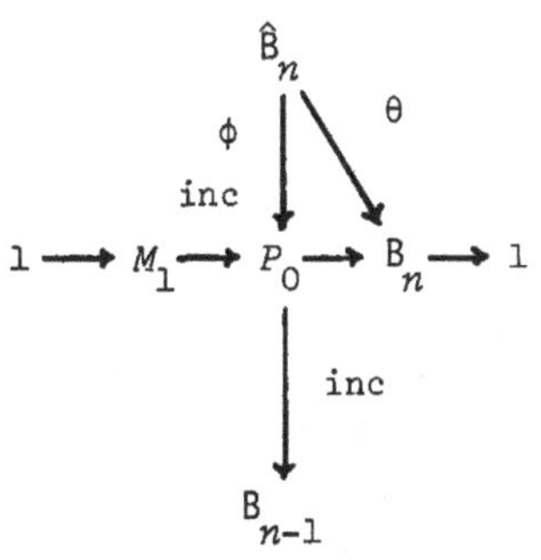

$$\langle a_0, \ldots, a_{n-1} | \rangle = F_n$$

where the row is split exact, P_0 is the group of 0-pure braids in B_{n+1}, and the map ϕ is easily written down. Under the assumption that ϕ is one-to-one (justified later) the problem of determining Ker θ reduces to that of describing (Ker θ)$\phi = M_1 \cap \text{Im}\,\phi$. This turns out to be the B_n-closure F of a single element $a_0^{-1}a_{n-1} \in F_n$, which is clearly just the subgroup consisting of all words of total exponent-sum zero. Hence, the programme is as follows:

(i) find Schreier generators for F,

(ii) compute the action of B_n on these induced by (2_n),

(iii) form the split extension $F \rtimes B_n$,

(iv) reduce the resulting presentation by Tietze transformations to that of $\hat{B}_n$.

This all goes smoothly except for (iv), which Albar [1] encompassed by developing a kind of "braid relation calculus", in which the crucial lemma may be of independent interest.

LEMMA. *If* y, d, x, t *are elements of any group, and* $y \approx dx, x, xd, t,$ *then*

$$y \approx t^{-1}x\,dx \Longleftrightarrow y \approx x\,dx\,t^{-1}.$$

THEOREM. $\hat{B}_n$ *is a split extension of a free group of countably infinite rank by* B_n.

4. The analogues

Any graph Γ with vertex set $1, 2, \ldots, n$ and edge set $\subseteq \{(i, j) \mid 1 \le i < j \le n\}$ determines a group G with generators $x_1, \ldots, x_n$ and defining relations

$$\left.\begin{array}{ll} x_i \approx x_j & \text{if there is an edge between } i \text{ and } j \\ x_i \sim x_j & \text{if there is no edge between } i \text{ and } j \end{array}\right\}.$$

Thus, for example, the graphs

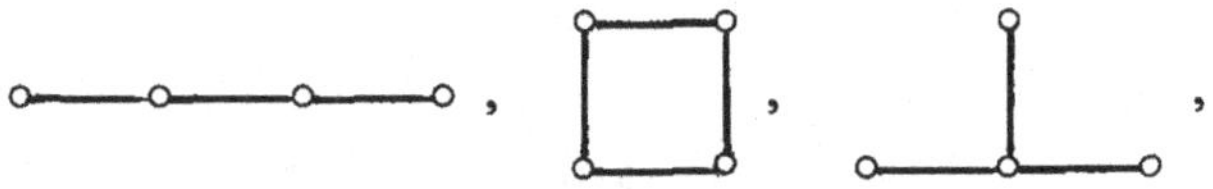

respectively determine B_5, $\hat{B}_4$, a new group, studied in more detail in [19]. It is only necessary to consider connected graphs, as components of Γ correspond to direct factors of G. As far as we can see at present, the structure of G depends heavily on that of Γ^*, the *complement* of Γ, which has an (i, j)-edge exactly where Γ doesn't. The following result ([19], [21]), for example, generalises known ones for the B_n.

THEOREM. *Suppose that* Γ *is connected and that* Γ^* *has at most two components. Then* G' *is finitely presented.*

References

[1] M. Albar, "On analogues of the braid group", Ph.D Thesis, Nottingham (1981).

[2] M. Albar and D.L. Johnson, "Circular braids", *Arab Gulf J. Sci. Research* 1, no. 2 (1983).

[3] K. Appel and P.E. Schupp, "Artin groups and infinite Coxeter groups", *Invent. Math.* 72 (1983), pp. 201-220.

[4] E. Artin, "Theorie der Zöpfe", *Abh. Math. Sem. Univ. Hamburg* 4 (1925), pp. 47-72, FdM.51, p. 450.

[5] J.S. Birman, *Braids, links and mapping class groups* (Ann. of Math. Studies 82, Princeton University Press, Princeton, 1974), MR51:4477, 54:13894.

[6] J.S. Birman and H.M. Hilden, "On isotopes of homeomorphisms of Riemann surfaces", *Ann. of Math.* (2) 97 (1973), pp. 424-439, MR48:4305.

[7] W.-L. Chow, "On the algebraical braid group", *Ann. of Math.* (2) 49 (1948), pp. 654-658, MR49, p. 98.

[8] H.S.M. Coxeter and W.O.J. Moser, *Generators and relations for discrete groups* (4th Edition, Ergebnisse der Mathematik 14, Springer-Verlag, Berlin-New York, 1980), MR81a:20001.

[9] F.A. Garside, "The braid group and other groups", *Quart. J. Math. Oxford* (2) 20, No. 78 (1969), pp. 235-254, MR40:2051.

[10] G.G. Gurzo, "The group of smooth braids", *16th All-Union Algebra Conference*, Abstract II, pp. 39-40, Leningrad (1981).

[11] D.L. Johnson, *Topics in the theory of group presentations* (Cambridge University Press, Cambridge, 1980), Zbl.437:20026.

[12] D.L. Johnson, "On a problem of Magnus", *J. Alg.* 79 (1982), pp. 121-126, Zbl.495:20017.

[13] D.L. Johnson, "Towards a characterization of smooth braids", *Math. Proc. Camb. Phil. Soc.* 92 (1982), pp. 425-427, MR84c:20049.

[14] D.L. Johnson and Lev J. Leifmann (eds.), *The Kourovka Notebook: Unsolved Problems in Theory* (Amer. Math. Soc. Translations (2) 121, 7th augmented edition, Providence, R.I., 1983).

[15] F. Laudenbach, "Présentation du groupe de difféotopies d'une surface compact orientable, Travaux de Thurston sur les surfaces", *Astérisque* 66-67 (1979), pp. 267-282, MR82m:57003.

[16] W. Magnus, "Braid groups: a survey", *Proc. Second Internat. Conf. on Theory of Groups*, Canberra, 1973 (Springer Lecture Notes 372, Berlin, 1974), pp. 463-487, MR50:5774.

[17] A.A. Markov, "Foundations of the algebraic theory of braids", *Trudy Mat. Inst. Steklov* 16 (1945), pp. 1-54 (Russian), MR8, p. 131.

[18] J. Mennicke, "Über Heegaarddiagramme vom Geschlect zwei mit endlicher

Fundamentalgrüppe", *Arch. Math.* 8 (1957), pp. 192-198, MR18:975.

[19] A.K. Napthine, "On the Artin braid group and some of its generalisations, Ph.D Thesis, University of Nottingham (1983).

[20] A.K. Napthine, "On some infinitely-generated subgroups of the braid groups", submitted to *J. London Math. Soc.*

[21] A.K. Napthine, "The derived groups of some generalisations of the braid groups", submitted to *Math. Proc. Camb. Philos. Soc.*

[22] B.H. Neumann and H. Neumann, "Embedding theorems for groups", *J. London Math. Soc.* (1) **34** (1959), pp. 465-497, MR29:1267.

[23] P.M. Neumann, "The *SQ*-universality of some finitely-presented groups", *J. Austral. Math. Soc.* **16** (1973), pp. 1-6, MR48:11342.

[24] S.J. Pride, "The concept of 'largeness' in group theory", *Word Problems, II (Conf. on Decision Problems in Algebra, Oxford, 1976)* (Studies in Logic and Foundations of Math. **95**, North-Holland, Amsterdam, 1980), pp. 229-335, MR81i:20038.

[25] W.P. Thurston and A.E. Hatcher, "A presentation for the mapping class group of a closed orientable surface", *Topology* **19**, no. 3 (1980), pp. 221-237, MR81k:57008.

Department of Mathematics,
University Park,
Nottingham NG7 2RD, England

PROCEEDINGS OF 'GROUPS — KOREA 1983'
KYOUNGJU, August 1983

20F32
20F38, 05C25

SOME USES OF COSET GRAPHS

R.C. Lyndon

Coset graphs have long been used in the study of permutation groups and, more recently, of fuchsian groups. (Stothers proposes to write a survey of their uses in connection with fuchsian groups.) Here we illustrate their use in some recent joint work with Brenner.

In 1933, Neumann [17] had occasion to study subgroups S of the modular group M that were maximal with respect to containing no parabolic element. He proved that if T is a maximal parabolic subgroup and S is a complement to T, then the *Neumann subgroup* S is maximal nonparabolic. A further study of these Neumann subgroups led Tretkoff [26] to a conjecture regarding their possible structure. Stothers [21], [22] preceded us [2], [3] in giving, independently, essentially the same proof of this conjecture. Magnus [14] raised the question of whether every maximal nonparabolic subgroup is a Neumann subgroup. We [4] obtained examples showing that this is not the case.

1. The structure of Neumann subgroups

The modular group M has a presentation

$$M = \langle a, b;\ a^2 = b^3 = 1\rangle .$$

The parabolic elements of M are precisely the conjugates of the nontrivial powers of the element $c = ab$.

Let S be any subgroup of M and let Γ be the coset graph of S relative to the two generators a, b of M. The vertices of Γ are the cosets Sg, $g \in M$. The set of (directed) edges e of Γ can be identified with the set $M \times \{a, b, b^{-1}\}$, with $e = (g, x)$ running from the vertex Sg to Sgx, and with inverse edge $e^{-1} = (gx, x^{-1})$. The edge $e = (g, x)$ has the *label* $\lambda(e) = x$. The a-orbits of the action of M on the vertices of Γ may be represented by the a-edges, joining a vertex Sg to Sga, where, possibly, $Sga = Sg$. The b-orbits may be represented either by a single b-edge joining Sg to itself if $Sgb = Sg$, or by a triangle of oriented b-edges joining Sg, Sgb, Sgb^2 cyclically.

Conversely, suppose a connected graph Γ is given with edges labelled by a, b, b^{-1}, and subject to the conditions above. Let S be the stabilizer in M of any vertex, under the obvious action of M on the vertices defined by the labelling of the edges. Then Γ is isomorphic to the coset graph of S.

By the Kurosh Subgroup Theorem,

$$S \simeq C_2^{*r_2} * C_3^{*r_3} * C_\infty^{*r_\infty}$$

the free product of r_2 groups of order 2, of r_3 groups of order 3, and of r_∞ infinite cyclic groups, where $0 \le r_2, r_3, r_\infty \le \infty$. Evidently

(1) r_2 is the number of fixed points of a in Γ;

(2) r_3 is the number of fixed points of b in Γ;

(3) the maximal nontrivial parabolic subgroups of S correspond bijectively to the finite c-orbits in Γ.

(A finite c-orbit $Sg, Sgc, \ldots, Sgc^{n-1}$, where $Sgc^n = Sg$, corresponds to the parabolic subgroup $\langle gc^n g^{-1}\rangle$.) Consequently

(3a) S is nonparabolic if and only if all c-orbits are infinite;

(3b) S is a Neumann subgroup if and only if Γ is infinite and c is transitive on the vertices of Γ.

The c-orbits of Γ induce on the quotient graph Γ/b an *eulerian system* Σ of reduced paths such that each (directed) edge of Γ/b occurs exactly once in a single path in Σ. From Γ/b and Σ we can recover the orientations of the b-orbits of Γ, and so reconstruct Γ. If S is a Neumann subgroup, then Σ reduces to a single *eulerian path* on Γ/b, and in this case we call Γ/b an *eulerian graph.*

It is technically convenient to modify Γ/b to a graph Γ^* by deleting all edges joining a vertex to itself. This leads to a mild ambiguity in reconstructing Γ that is harmless for our purposes. The graph Γ^* is now *cuboid*: there are at most 3 edges at each vertex; the paths of the eulerian system are now reduced (without U-turns) except, necessarily, at vertices of degree 1.

Analysis shows that any connected eulerian cuboid graph Γ^* is obtainable from a eulerian cubic graph Γ_0^* by attaching trees; at most one of these trees can be infinite, and in that case it is *simply infinite*, with exactly one infinite reduced path leading out of each vertex.

From this information it is not difficult to deduce the free product structure of S, and to show that r_∞ is the Betti number of Γ^*, and is always even when it is finite. (There is here a connection with the coinitial graph of Hoare, Karrass, and Solitar [11], [12].) Moreover, every choice of the parameters r_2, r_3, r_∞ subject to these conditions arises for some Neumann subgroup S . This confirms Tretkoff's conjecture.

2. Maximal nonparabolic groups that are not Neumann subgroups

Let N be the kernel of the obvious map from M onto the group

$$Q = \langle a, b;\ a^2 = b^3 = (abab^{-1})^3 = 1\rangle$$

of Coxeter [9], [10] and Sinkov [18]. It will be shown that N is maximal in the class of normal nonparabolic subgroups of M , and, more important for our purpose, that N is contained in no Neumann subgroup.

The group Q arises as a subgroup of index 2 in the full symmetry group of the regular tessellation H of the euclidean plane by hexagons. The *truncation* H^* of H is obtained by replacing each vertex of H , with its 3 incident edges, by a small circle C , the 3 edges now terminating on C . Now H^* serves as a coset graph for N , the a-edges being the straight edges and the b-orbits running around the small circles, alternately in the positive and the negative sense. The c-orbits of H fall into 3 infinite families K, L, M , each consisting of infinitely many parallel orbits. See Figure 1.

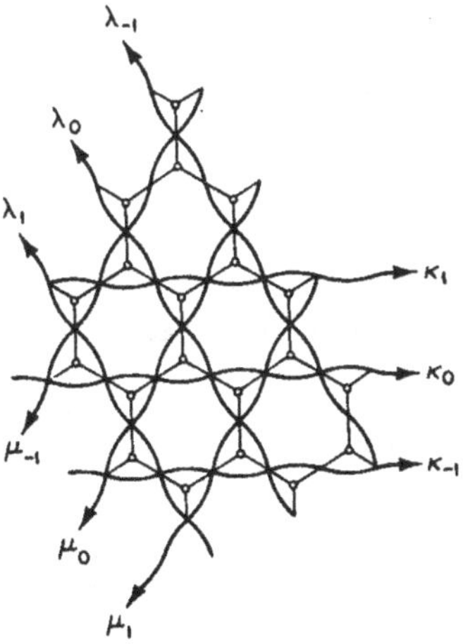

Figure 1.

The positively oriented circles C correspond bijectively to the triples (κ, λ, μ) in $K \times L \times M$, with C incident with κ, λ, μ . We represent H^* by the regular tessellation T of the Euclidean plane, with the vertices of T representing the positively oriented circles and the infinite straight lines in T representing the c-orbits. A group S , with $N \le S \le M$, corresponds to a quotient graph of

H^* , hence to a quotient graph Δ of T , and, in fact, to one of the form $\Delta = T/G$ for G a group of rigid motions of T . Moreover, S will be nonparabolic if and only if G contains no nontrivial translation in the direction of any c-orbit. This restricts G to be trivial, a rotation group of order 3 , an infinite cyclic translation group, or one of the associated dihedral groups.

It is not difficult to single out those G for which S , associated with $\Delta = T/G$, is maximal nonparabolic, and to verify that in all cases there is more than one c-orbit. The simplest case is given by G infinite dihedral, with a translation of minimal length. The coset graph for S is shown in Figure 2, from which it can be read off that S is the free product of groups S_n of order 2 , for all n in Z , where $S_n = \langle c^n bab^{-1}c^{-n} \rangle$.

Figure 2.

3. Complementary remarks

(3.1) Tretkoff obtained from Γ the following presentation for S :

$$S = \langle V;\ \{p(pa) = 1\},\ \{p(pb)(pb^2) = 1\}\ ,\quad \text{all}\ \ p \in V \rangle\ ,$$

where V is the set of vertices of Γ . Clearly any set of permutations $a, b, \ldots$ of a set V determines a group presentation in a similar manner. In the case that Γ has no fixed points and a cyclic order of the edges at each vertex is given, we have used a variant of Tretkoff's presentation:

$$S = \langle E;\ \{r_e = 1\},\ \{r_v = 1\},\ e \in E,\ v \in V \rangle\ ,$$

where E is the set of edges of Γ , $r_e = ee^{-1}$, and $r_v = e_1 \ldots e_n$, the product of the edges at v in the given cyclic order. Note that, for any graph Γ , the latter presentation is quadratic in the sense of Hoare, Karrass, and Solitar.

(3.2) In [3] a modest catalog of eulerian cuboid graphs Γ^* with few vertices led us to conjecture that the automorphism group of Γ^* , necessarily cyclic, must have order 1 , 2 , 3 , or 6 . Bianchi and Cori [1] proved this, indeed with 2 and 3 replaced by any pair of distinct primes.

(3.3) In [6] we investigated infinite eulerian graphs in the plane, and showed that the 1-skeleton of every regular tessellation of the euclidean or hyperbolic plane is eulerian. With a small exception, we proved the same for irregular tessellations of the plane that satisfy a 'small cancellation condition' $C(p,q)$. (Here it is suggestive that a lemma from small cancellation theory enters essentially into our proof.) In [7] we show that the rectangular grid in dimensions $d \geq 2$ is eulerian.

(3.4) In [5] we use graph theoretic methods to study other triples a, b, $c = ab$ of permutations, each subject to a condition on its order or the condition of transitivity. For example, given any a, provided the lengths of its orbits satisfy a certain rather obvious inequality, there exists b such that $b^2 = 1$ and $c = ab$ is transitive. This was proved by Luoto [13], who also showed that Γ can be embedded in the euclidean plane with all a-edges and b-edges straight segments of the same length. We prove also a variant of a theorem of Miller [15], [16], that, given integers $\alpha, \beta, \gamma \geq 2$, there exist permutations $a, b, c = ab$ of a finite set Ω, of orders α, β, γ. In [8] we sharpen Miller's result concerning the minimum cardinality $d(\alpha, \beta, \gamma)$ of Ω.

Acknowledgement

The author gratefully acknowledges partial support of this research by the National Science Foundation.

References

[1] G. Bianchi and R. Cori, "Colorings of hypermaps and a conjecture of Brenner and Lyndon", *Pacific J. Math.* 110 (1984), pp. 41-48.

[2] J.L. Brenner and R.C. Lyndon, "Nonparabolic subgroups of the modular group", *J. Algebra* 77 (1982), pp. 311-322, MR83k:10048.

[3] J.L. Brenner and R.C. Lyndon, "Permutations and cubic graphs", *Pacific J. Math.* 104 (1983), pp. 285-315.

[4] J.L. Brenner and R.C. Lyndon, "Maximal nonparabolic subgroups of the modular group", *Math. Ann.* 263 (1983), pp. 1-11.

[5] J.L. Brenner and R.C. Lyndon, "The orbits of a product of two permutations", *European J. Combin.*, to appear.

[6] J.L. Brenner and R.C. Lyndon, "Infinite Eulerian tessellations", *Discrete Math.* 46 (1983), pp. 111-132.

[7] J.L. Brenner and R.C. Lyndon, "Doubly Eulerian trails on rectangular grids", *J. Graph Theory*, to appear.

[8] J.L. Brenner ans R.C. Lyndon, "A theorem of G.A. Miller on the order of the product of two permutations. I": *Jnanabha*, to appear; II: *Indian J. Math.*, to appear; III: *Pure Appl. Math. Sci.*, to appear.

[9] H.S.M. Coxeter, "The groups determined by the relations $S^l = T^m = (S^{-1}T^{-1}ST)^p = 1$", *Duke Math. J.* 2 (1936), pp. 61-73, FdM62, p. 80.

[10] H.S.M. Coxeter and W.O.J. Moser, *Generators and relations for discrete groups* (editors, Ergebnisse der Mathematik, 14, Berlin, Heidelberg, New York, 1972), MR50:2313.

[11] A.H.M. Hoare, A. Karrass, and D. Solitar, "Subgroups of finite index of Fuchsian groups", *Math. Z.* 120 (1971), pp. 289-298, MR44:2837.

[12] A.H.M. Hoare, A. Karrass, and D. Solitar, "Subgroups of infinite index in Fuchsian groups", *Math. Z.* 125 (1972), pp. 59-69, MR45:2029.

[13] K. Luoto, Private communication.

[14] W. Magnus, *Noneuclidean tesselations and their groups* (Academic Press, London, New York, 1974), MR50:4774.

[15] G.A. Miller, "On the product of two substitutions", *Amer. J. Math.* 22 (1900), pp. 185-190, FdM31, p. 134.

[16] G.A. Miller, "Groups defined by the orders of two generators and the order of their product", *Amer. J. Math.* 24 (1902), pp. 96-100, FdM33, p. 154.

[17] B.H. Neumann, "Über ein gruppentheoretisch-arithmetisches Problem", *Sitzungsber. Preuss. Akad. Wiss. Phys. Math. Kl.* No. 10 (1933), pp. 429-444, FdM59, p. 146.

[18] A. Sinkov, "The groups determined by the relations $S^l = T^m = (S^{-1}T^{-1}ST)^p = 1$", *Duke Math. J.* 2 (1936), pp. 74-83, FdM62, p. 80.

[19] W.W. Stothers, "Subgroups of the modular group", *Proc. Camb. Phil. Soc.* 75 (1974), pp. 139-153, MR48:10988.

[20] W.W. Stothers, "Subgroups of the (2, 3, 7) triangle group", *Manuscripta Math.* 20 (1977), pp. 323-334, MR56:2923.

[21] W.W. Stothers, "Subgroups of infinite index in the modular group", *Glasgow Math. J.* 19 (1978), pp. 33-43, MR80b:20060.

[22] W.W. Stothers, "Diagrams associated with subgroups of Fuchsian groups", *Glasgow Math. J.* 20 (1979), pp. 103-114, MR80j:20048.

[23] W.W. Stothers, "Subgroups of infinite index in the modular group, II", *Glasgow Math. J.* 22 (1981), pp. 101-118, MR83m:10033a.

[24] W.W. Stothers, "Subgroups of infinite index in the modular group, III", *Glasgow Math. J.* 22 (1981), pp. 119-131, MR83m:10033b.

[25] W.W. Stothers, "Groups of the second kind with the modular group, III", *Illinois J. Math.* 25 (1981), pp. 390-397, MR83a:10041.

[26] C. Tretkoff, "Non-parabolic subgroups of the modular group", *Glasgow Math. J.* 16 (1975), pp. 91-102, MR53:3140.

Department of Mathematics,
University of Michigan,
Ann Arbor, Michigan 48109
U.S.A.

PROCEEDINGS OF 'GROUPS — KOREA 1983'
KYOUNGJU, August 1983

20H05,
20H25

DISCONTINUOUS GROUPS

J.L. Mennicke

This is a resumé of a survey talk given at the Conference Groups, Korea (1983).

1) We shall be interested in groups acting discontinuously on some symmetric domain. Here is a standard example:

$$\Gamma = \mathrm{SL}_n(\mathbb{Z}) .$$

We shall be interested in the normal structure of Γ, i.e. we ask for a complete survey of the normal subgroups of Γ.

There are certain intrinsic normal subgroups:

a) $$N_m = \{x \in \Gamma, \quad x \equiv I \bmod m\} \quad \text{for} \quad m \in \mathbb{N} .$$

One can easily prove

$$N_m \lhd \Gamma \quad \text{and} \quad \Gamma/N_m \cong \mathrm{SL}_n(\mathbb{Z}/m\mathbb{Z}) .$$

b) Central normal subgroups:

$$U \leq \{\lambda I, \ \lambda = \pm 1\} .$$

Here is the structure theory for normal subgroups of Γ.

THEOREM 1. *1) Suppose $n \geq 3$. Let $N \lhd \Gamma$ be any normal subgroup in Γ. Then either*

(i) there is an $m \in \mathbb{N}$ such that $N_m < N$, or

(ii) N is a central normal subgroup.

2) Suppose $n = 2$. Then Γ has many normal subgroups.

We shall outline some generalizations. Let $k/\mathbb{Q}$ be an algebraic number field, and let $o \subset k$ be the ring of algebraic integers in k. Consider

$$\Gamma = \mathrm{SL}_n(o) .$$

THEOREM 2. *Consider the two alternatives for a normal subgroup N, either* *(i)* $N \lhd \Gamma$ *is of finite index, or*

(ii) N *is a central normal subgroup.*

1) *Suppose* $n \geq 3$. *Then the above alternatives hold.*

2) *Suppose* $n = 2$. *Then the alternatives hold if* k *contains infinitely many units, i.e. if we exclude*

a) $k = \mathbb{Q}$, $o = \mathbb{Z}$

b) $k = \mathbb{Q}(\sqrt{-N})$, $N \in \mathbb{N}$, N *square-free* .

We shall further generalize. Let k be an algebraic number field. Let G be an algebraic group defined over k . Let $G(k)$ be the group of k-points of G .

Assume that G is semi-simple.

Let S be a finite set of valuations of k , including the Archimedian valuations. Let o_S be the ring of numbers of k which are integral for all valuations not S .

THEOREM 3. *In the above notation consider the group* $\Gamma = G(o_S)$.

Under some mild assumptions pertaining to the k*-rank of* $G(k)$ *and to the cardinality of the set* S *, the alternatives hold:*

Any normal subgroup $N \lhd \Gamma$ *is either of finite index or central.*

Notice that Theorems 2 and 3 are qualitative in nature. In certain special cases, quantitative versions are available. In the general situation of Theorem 3, a full quantitative theory is not yet available. The difficulties which so far have not been overcome are of a different nature. In some cases, e.g. for the orthogonal groups and their covering groups, a number of prime divisors are special in the given context, e.g. the divisors of the discriminant. These special divisors would produce deviations from a congruence subgroup theorem such as Theorem 1. A more serious difficulty arises in the absence of unipotent elements in Γ . This difficulty has not been overcome in any special case.

We shall not specify the conditions mentioned in Theorem 3. Rather we shall give examples where they hold and where they do not hold.

EXAMPLE 1. *Consider the quadratic form*

$$f_1 = x_1x_2 + x_3x_4 + \dots + x_{2n-1}x_{2n} , \quad n \geq 2 ,$$

and the group

$$\Gamma = SO_{2n}(f_1, \mathbb{Z}) .$$

For this group, the alternatives of Theorem 3 hold.

EXAMPLE 2. *Consider*

$$f_2 = -x_1^2 - x_2^2 - \dots - x_{n-1}^2 + x_n^2 , \quad n \geq 2 .$$

For the groups $\Gamma = SO_n(f_2, \mathbb{Z})$, the alternatives of Theorem 3 do not hold.

We offer a few historical comments. Theorem 1 was proved independently by Bass-Lazard-Serre [1] and by the author. A good account of Theorem 2 can be found in Bass-Milnor-Serre. The very powerful Theorem 3 is the work of G.A. Margulis. There is a survey due to Tits [8].

2) We shall now describe some work concerned with the cases excluded under 2b) in Theorem 2.

Hence we deal with

$k = \mathbb{Q}(\sqrt{-N})$, $N \in \mathbb{N}$ square-free , o the ring of integers in k ,

$\Gamma = PSL_2(o)$.

In cases where the sources are not explicitly quoted, the results are joint work with Elstrodt, Grunewald and the author.

There are a number of qualitative results which we shall state more or less in historical order.

THEOREM 4 (Borel, Harish Chandra). *The groups* Γ *described above are all finitely presented.*

This is, of course, a very special instance of a very general theorem.

THEOREM 5 (Serre). *Consider the rank of the commutator quotient group of* $\Gamma = PSL_2(o)$. *Then*

$$\mathrm{rk}(\Gamma^{ab}) \geq h - 1 .$$

Here h *denotes the class number of* $k = \mathbb{Q}(\sqrt{-N})$.

The proof is based on a simple topological argument containing 3-manifolds.

THEOREM 6 (Zimmert). *(i) There is an integer* $L \in \mathbb{N}$ *, depending on the discriminant of* k *, and* $L > 1$ *for all sufficiently large* N *such that* Γ *has a free non-abelian quotient group* F_L *of rank* L .

(ii) The estimate in Theorem 5 can be improved to

$$\mathrm{rk}(\Gamma^{ab}) \geq h - 1 + L .$$

The proof depends on an ingenious topological construction which has been applied successfully also in other situations. The method may still have other applications.

Using advanced number theory and invoking the Generalized Riemann Hypothesis, Grunewald and Zimmert were able to show $L \to \infty$ as $N \to \infty$.

THEOREM 7 (Rohlfs). *There is an integer* L_1 *, depending on* N *, and* > 1 *for all sufficiently large* N *, such that*

$$\mathrm{rk}(\Gamma^{ab}) \geq k - 1 + L_1 .$$

$L_1 \to \infty$ *as* $L \to \infty$ *, without invoking any hypothesis.*

The number L_1 in Theorem 7 is of the same order of magnitude as L in Theorem 6. The proof of Theorem 7 is based on a version of the Lefschetz fixed point formula which was developed for this purpose.

Notice that Theorems 5, 6, 7 are of a qualitative nature. Extensive numerical computations have revealed that the groups treated in this section are tied-up with very delicate arithmetical questions. We shall discuss these questions in the remaining part of this survey, in a very special instance. It is reasonable to describe the observations as Grunewald's observations.

Consider the special case $k = \mathbb{Q}(\sqrt{-1})$, $o = \mathbb{Z}[i]$. Consider a prime ideal $p \in \mathbb{Z}[i]$ of degree one, i.e. $Np = \mathrm{p}$, where $\mathrm{p} \equiv 1 \bmod 4$, and the Hecke subgroups

$$\Gamma_o(p) = \left\{X = \begin{pmatrix} \alpha & \beta \\ \gamma & \delta \end{pmatrix} \in \Gamma \, , \, \gamma \in p\right\} .$$

We have carried out extensive computations for $\Gamma_o(p)^{ab}$ and its rank r . Here is a small proportion of our results:

p	r
5	0
13	0
⋮	
137	1
---	0
233	1
---	0
257	1
---	0
277	1
---	-

It has become clear that this very exotic distribution cannot be explained by abelian class field theory. Grunewald has found a very beautiful numerical relationship to elliptic curves.

Grunewald's observations: In every $r = 1$-case, for all $\mathrm{p} < 2000$, there is an elliptic curve $E/\mathbb{Q}(i)$, unique up to isogeny, such that E has good reduction for all $q \neq p$, $(1+i)$, and conductor conditions for the excluded primes which amount to multiplicative reduction.

The relationship is established analytically as follows: The groups $\Gamma_o(p)^{ab}$ come with an intrinsic algebra of $\mathbb{Z}$-endomorphisms, known as the Hecke-algebra, which can be defined purely algebraically, and produces eigenvalues which are rational integers. Use these eigenvalues to define an L-function, via an Euler product. Invoking

A. Weil's "Converse Theorem", one can show that this *L*-function is meromorphic in the entire complex plane, and comes with a functional equation. The *L*-function coincides with the Hasse-Weil ζ-function $\zeta(E)$.

Coincide means, in our context, that the first few hundred Euler factors on either side coincide.

Observe that Grunewald's observation is of a heuristical nature, based on extensive numerical computations. There is no reason to doubt the significance of the observation. There are a few cases for small values of r other than 1 . In these cases, there are analogous versions of Grunewald's observation.

There is one more link between $\Gamma_o(p)$ and elliptic curves.

THEOREM 8. *In the cases* $p = 137$ *and* $p = 257$ *, the elliptic curves have* k*-rational points of order* 2 *and* 4 *, respectively.*

For the eigenvalues of the Hecke algebra, there are congruences mod 2 and mod 4 , respectively. These congruences can be proved.

Observe that Theorem 8 would follow, once the coincidence between $L(S)$ and $\zeta(E)$ were established.

Grunewald's observation is not restricted to the gaussian integers. It has analogues in other number fields of small discriminant. The following speculation has not yet reached the more dignified status of a conjecture:

$\mathrm{rk}(\Gamma^{ab})$ and the ranks of subgroups of finite index which are congruence subgroups should be controlled by abelian varieties over k with pre-assigned reduction properties.

To conclude this survey talk, we offer a few remarks concerning the group theoretic structure of groups $\Gamma = \mathrm{PSL}_2(o)$. Computations of many special instances have shown that these groups and some of their natural extensions known as extended Bianchi groups come with very complicated presentations. In some cases, Waldhausen's theory of 3-manifolds with sufficiently large fundamental groups is applicable. The result is a description of Γ by a chain of HNN-extensions and free products with amalgamations. It is not clear that almost all fundamental domains contain incompressible surfaces and hence are accessible to the Haken-Waldhausen methods. Moreover, it is not clear whether or not the occurring HNN-extensions and free products with amalgamations are restricted by the fact that we are working with algebraic groups or whether they are of a general nature. There are hints that the latter might be true.

It is my pleasure to offer thanks to the organizer of the Conference for excellent hospitality. Thanks are also due to Deutsche Forschungsgemeinschaft for financial support.

References

[1] H. Bass, J. Milnor, and J.-P. Serre, "Solution of the congruence subgroup problem for $SL_n (n \geq 3)$ and $Sp_{2n} (n \geq 2)$", *Publ. Math. I.H.E.S.* 33 (1967), pp. 59-137, MR39:5574.

[2] J. Elstrodt, F. Grunewald, and J. Mennicke, "On the group $PSL_2(\mathbb{Z}[i])$", *Proc. Journées Arithmétiques* 1980 (LMS Lectures 56, London, 1982), pp. 255-283.

[3] J. Elstrodt, F. Grunewald, and J. Mennicke, "Discontinuous groups on 3-dimensional hyperbolic space: Analytic theory and number theoretic applications", *Uspekhi Mat. Nauk* 38:1 (1983), no. 229, pp. 119-147.

[4] J. Elstrodt, F. Grunewald, and J. Mennicke, "PSL(2) over imaginary quadratic integers", *Journées Arithmétiques Metz* (1981), *Astérisque* 94 (1982), pp. 43-60.

[5] J. Mennicke, "Finite factor groups of the unimodular group", *Annals of Mathematics* (2) 81 (1965), pp. 31-37, MR30:2083.

[6] J. Rohlfs, "On the cuspidal cohomology of the Bianchi modular groups", Preprint Eichstätt (1983).

[7] J.-P. Serre, "Le problème des groupes de congruence pour SL_2", *Annals of Mathematics* (2) 92 (1970), pp. 489-527, MR42:7671.

[8] J. Tits, "Travaux de Margulis sur les sous-groupes discrets de Lie", Séminaire Bourbaki 28ème année (1975/1976), Exp. 482 (Lectures Notes in Mathematics 567, Springer, Berlin, 1977), MR58:11227.

[9] R. Zimmert, "Zur SL_2 der ganzen Zahlen eines imaginär-quadratischen Zahlkörpers", *Inventiones Mathematicae* 19 (1973), pp. 73-81, MR47:6883.

Fakultät für Mathematik,
Universität Bielefeld,
4800 Bielefeld 1,
Federal Republic of Germany

PROCEEDINGS OF 'GROUPS — KOREA 1983'
KYOUNGJU, August 1983

20N99

COMMUTATIVE QUANDLES

Dedicated to Shin-ichi Izumi for his 80th Birthday, 1984-07-02

B.H. Neumann

David Joyce [1], in pursuit of a classifying invariant of knots, introduced the term "quandle" for an algebraic system with two binary operations; one of them mimics conjugations in a group, the other undoes the effect of the first. Similar algebraic systems had occurred in the literature before: see Joyce [1] for a discussion and references. In a second paper [2], Joyce further investigates quandles, and especially the finite simple quandles (that is those with only the obvious congruences: the trivial congruence and the universal congruence). I here adopt a different notation from that of Joyce.

In a group, define binary operations σ and τ by

$$xy\sigma = y^{-1}xy$$

$$xy\tau = yxy^{-1} .$$

Then σ , τ satisfy the laws

Q1: $$xx\sigma = x ,$$

Q2: $xy\sigma y\tau = x$, Q2': $xy\tau y\sigma = x$,

Q3: $$xy\sigma z\sigma = xz\sigma yz\sigma\sigma ,$$

and their consequences. Define a *quandle* as an algebraic system with two binary operations σ , τ , subject to these laws.

Groups, with the stated ("standard") interpretation of σ and τ , are obvious examples of quandles; but one need not look at a whole group: thus a conjugacy class in a group, with the standard interpretation of σ and τ , becomes a quandle.

EXAMPLE 1.

σ	0	1	2
0	0	2	1
1	2	1	0
2	1	0	2

τ	0	1	2
0	0	2	1
1	2	1	0
2	1	0	2

This exhibits the conjugacy class of the 3 elements of order 2 in the symmetric group of degree 3. Here $\tau = \sigma$, which is equivalent to the "involutory" law

$$xy\sigma y\sigma = x .$$

From the quandle laws Q1, Q2, Q2', Q3, other laws follow, for example the "dual" laws

Q1': $$xx\tau = x ,$$

Q3': $$xy\tau z\tau = xz\tau yz\tau\tau ,$$

and "mixed" laws

Q4: $$xy\sigma z\tau = xz\tau yz\tau\sigma ,$$

Q4': $$xy\tau z\sigma = xz\sigma yz\sigma\tau .$$

A quandle that arises from a set of elements of a group with the standard interpretation of σ and τ will here be called *representable*. Not all quandles are representable. In a representable quandle the following identical implication is valid:

$$\text{if} \quad xy\sigma = x , \quad \text{then} \quad yx\sigma = y .$$

This does not follow from the quandle laws.

EXAMPLE 2.

σ	0	1	2
0	0	0	0
1	2	1	1
2	1	2	2

τ	0	1	2
0	0	0	0
1	2	1	1
2	1	2	2

It is not difficult to verify that the quandle laws are satisfied, but the above identical implication is not.

The quandle operations naturally lead to the definition of mappings of the carrier (or set of elements) of a quandle into itself, two to each element y of the quandle, namely S_y and T_y defined by

$$xS_y = xy\sigma , \qquad xT_y = xy\tau .$$

The quandle laws then translate to

S1: $$xS_x = x ;$$

S2: $$S_yT_y = I ,$$ S2': $$T_yS_y = I ,$$

where I is the identity mapping of the carrier;

S3: $$S_y S_z = S_z S_{yz\sigma} .$$

Thus all S_y and T_y are permutations of the carrier, and $T_y = S_y^{-1}$. They generate a permutation group, which I call simply the *quandle group* (Joyce [2] calls it the *inner automorphism group* of the quandle). The law S3 and the law that corresponds to Q4, namely

S4: $$S_y T_z = T_z S_{yz\tau} ,$$

guarantee that the generators S_y of the quandle group all lie in a single conjugacy class of the group. Consequently, if W is a set of group words in "variables", then the values of all $w \in W$, with generators S_y substituted for the variables, generate a normal subgroup of the quandle group. Joyce [2] notes this in the case where W consists of the single word $w = yz^{-1}$, and calls the resulting normal subgroup, generated by all $S_y S_z^{-1}$, the *transvection group* of the quandle. He then calls the quandle *abelian* if this transvection group is abelian. This is equivalent to imposing the law

QAB: $$xy\sigma zt\sigma\sigma = xz\sigma yt\sigma\sigma$$

on the quandle; or, again equivalently, the dual law

QAB': $$xy\tau zt\tau\tau = xz\tau yt\tau\tau .$$

The proof of the equivalence of QAB and QAB' is easy and omitted. To distinguish this kind of quandle from the commutative quandles, to be introduced presently, I shall call them *Joyce abelian*. The two examples given above are both Joyce abelian.

EXAMPLE 3.

σ	0	1	2	3	4
0	0	3	1	4	2
1	3	1	4	2	0
2	1	4	2	0	3
3	4	2	0	3	1
4	2	0	3	1	4

τ	0	1	2	3	4
0	0	4	3	2	1
1	2	1	0	4	3
2	4	3	2	1	0
3	1	0	4	3	2
4	3	2	1	0	4

This is also a Joyce abelian quandle, but unlike Examples 1 and 2 it is not involutory. In it

$$01\sigma 23\tau\sigma = 2 \neq 0 = 02\sigma 13\tau\sigma ,$$

showing that QAB does not imply a "mixed" law.

Example 3 also exhibits what I call *σ-symmetry* :

$$xy\sigma = yx\sigma ,$$

because the σ table is symmetric about the main diagonal. However, as the example shows, σ-symmetry does not imply τ-symmetry, and it will not be further considered here.

Instead I introduce the law

QC: $$xy\sigma z\sigma = xz\sigma y\sigma ,$$

which translates to

SC: $$S_y S_z = S_z S_y ,$$

and thus implies, and is implied by, the commutative law for the quandle group. I call a quandle with this property *commutative*. Again the dual commutative law

QC': $$xy\tau z\tau = xz\tau y\tau$$

can be deduced from QC, and also the "mixed" law

QD: $$xy\sigma z\tau = xz\tau y\sigma ,$$

which is self-dual. Using the law Q3, one derives from QC the "absorptive" law

QE: $$xyz\sigma\sigma = xy\sigma .$$

This is, in turn, equivalent to its dual

QE': $$xyz\tau\tau = xy\tau ,$$

and to the mixed laws

$$xyz\tau\sigma = xy\sigma ,$$

$$xyz\sigma\tau = xy\tau .$$

Conversely, QC follows from QE: they are equivalent. The easy proof is omitted. The law QAB is an easy consequence of QE and QC:

$$xy\sigma zt\sigma\sigma = xy\sigma z\sigma = xz\sigma y\sigma = xz\sigma yt\sigma\sigma .$$

Thus every commutative quandle is Joyce abelian. The converse is not true: the quandles of all three examples above are Joyce abelian, but only the one of Example 2 is commutative. Indeed the quandle group of Example 1 is the symmetric group of degree 3, that of Example 2 is $\mathbb{Z}_3$, the cyclic group of order 3, and that of Example 3 is the (necessarily split) extension of $\mathbb{Z}_5$ by its full automorphism group $\mathbb{Z}_4$, also known as the holomorph of $\mathbb{Z}_5$. If however, the whole of a group is turned into a quandle with the standard interpretation of the operations, then the quandle is Joyce abelian (if and) only if it is commutative, namely if and only if the group to start with was nilpotent of class at most 2.

It is possible to give a "normal form" for the elements of a finitely generated commutative quandle. First of all the elements of the free commutative quandle on n generators can be represented by $(n+1)$-vectors in $\mathbb{Z}^{n+1}$,

$$q = (i, a_1, a_2, \ldots, a_n)$$

subject to

$$1 \leq i \leq n, \qquad a_i = 0 .$$

If

$$r = (j, b_1, b_2, \ldots, b_n)$$

is another such vector, then

$$qr\sigma = (i, c_1, c_2 \ldots, c_n) ,$$

where, if $j = i$, then $c_\nu = a_\nu$ for $1 \leq \nu \leq n$, and if $j \neq i$, then $c_j = a_j + 1$, and $c_\nu = a_\nu$ for $1 \leq \nu \leq n$, $\nu \neq j$. Similarly

$$qr\tau = (i, d_1, d_2, \ldots, d_n) ,$$

where, if $j = i$, then $d_\nu = a_\nu$ for $1 \leq \nu \leq n$, and if $j \neq i$, then $d_j = a_j - 1$, and $d_\nu = a_\nu$ for $1 \leq \nu \leq n$, $\nu \neq j$. More concisely,

$$qr\sigma = (i, a_1, \ldots, a_{j-1}, a_j + (1 - \delta_{ij}), a_{j+1}, \ldots, a_n) ,$$

$$qr\tau = (i, a_1, \ldots, a_{j-1}, a_j - (1 - \delta_{ij}), a_{j+1}, \ldots, a_n) ,$$

where δ_{ij} is the Kronecker delta. The quandle group is the free abelian group $\mathbb{Z}^{n(n-1)}$ of rank $n(n-1)$. The free generators are the vectors

$$g_i = (i, 0, 0, \ldots, 0) , \qquad 1 \leq i \leq n .$$

When relations are introduced into the free commutative quandle, the situation becomes more complex, and will not be described here. An example must suffice. Let κ be a congruence on the free commutative quandle on two generators

$$g_1 = (1, 0, 0) , \qquad g_2 = (2, 0, 0) .$$

Assume that κ is non-trivial, so that there is at least one pair $(q, r) \in \kappa$ with $q \neq r$; write $q \equiv r$ for simplicity, κ being understood. Then four cases arise.

I. If $q = (1, 0, a_2)$ and $r = (2, b_1, 0)$ in the above notation then $g_1 q\sigma \equiv g_1 r\sigma$, that is

$$(1, 0, 0) \equiv (1, 0, 1) .$$

Applying $g_2\sigma$ or $g_2\tau$ repeatedly on the right shows that all $(1, 0, a)$ are congruent to each other. Symmetrically, all $(2, b, 0)$ are congruent to each other.

Finally, as $(1, 0, a_2) \equiv (2, b_1, 0)$, all elements are congruent to each other, and κ is the universal congruence.

II. If $q = (1, 0, a_2)$, $r = (1, 0, b_2)$, then repeated application of $g_2\sigma$ or $g_2\tau$ on the right gives

$$(1, 0, a) \equiv (1, 0, a') \quad \text{whenever} \quad a \equiv a' \bmod c ,$$

where $c = |b_2 - a_2|$. Thus the elements of the resulting quotient quandle can be represented by triplets $(1, 0, a)$ with $a \in \mathbb{Z}_c$ and $(2, b, 0)$ with $b \in \mathbb{Z}$. In this case the quandle group is $\mathbb{Z}_c \times \mathbb{Z}$.

III. If $q = (2, a_1, 0)$, $r = (2, b_1, 0)$, then the symmetrical argument gives

$$(2, a, 0) \equiv (2, a', 0) \quad \text{whenever} \quad a \equiv a' \bmod d ,$$

where $d = |b_1 - a_1|$. The elements of the quotient quandle are represented by $(1, 0, a)$ with $a \in \mathbb{Z}$, and $(2, b, 0)$ with $b \in \mathbb{Z}_d$.

IV. If κ contains pairs of both kinds above, then the resulting quotient quandle becomes finite. Its elements are triplets of the form $(1, 0, a)$ with $a \in \mathbb{Z}_c$ and $(2, b, 0)$ with $b \in \mathbb{Z}_d$, for some positive integers c, d , and the order of the quandle is $c + d$. The quandle group in this case is $\mathbb{Z}_c \times \mathbb{Z}_d$. Example 2 exhibits this quandle when $c = 1$ and $d = 2$. A quandle like this is representable if and only if $c = d$; the proof, which is not difficult, is omitted.

References

[1] David Joyce, "A classifying invariant of knots, the knot quandle", *J. Pure Appl. Algebra* 23 (1982), pp. 37-65, MR83m:57007.

[2] David Joyce, "Simple quandles", *J. Algebra* 79 (1982), pp. 307-318, MR84d:20078.

Department of Mathematics,
Institute of Advanced Studies,
Australian National University,
Canberra, A.C.T., Australia,
and
Division of Mathematics and Statistics,
Commonwealth Scientific and Industrial Research Organization,
Canberra, A.C.T., Australia.

PROCEEDINGS OF 'GROUPS — KOREA 1983' 20D10, 20D15, 20J04
KYOUNGJU, August 1983

METABELIAN GROUPS OF PRIME-POWER EXPONENT

M.F. Newman

Abstract

Some structural features of relatively free metabelian groups of prime-power exponent are described. Earlier results are surveyed. New results depend critically on the use of a computer implementation of the nilpotent quotient algorithm modified to take advantage of the metabelian context.

1. Introduction

The questions Burnside asked about periodic groups in 1902 and related questions which have been asked since have been remarkably fruitful in leading to a better understanding of these groups. At present the most challenging problem in this area appears to be: among finite d-generator groups of prime-power exponent, p^k say, is there a largest one? (A discussion of such problems can be found in Newman [19].) The answer is of course yes whenever the answer is yes in the corresponding case for Burnside's question; that is, for $p^k = 2, 3$ or 4. In addition Kostrikin [13], [14], [15] has given the answer yes in the prime case $(k = 1)$. Otherwise the problem is almost completely open. A beginning has been made for groups of exponent 8 (Alford and Pietsch [1]; Havas and Newman [10], p.226).

Kostrikin's work is based on the observation that for elements a, b of a group G of exponent p the iterated left-normed commutator $[b, \underbrace{a, \ldots, a}_{p-1}]$, or briefly $[b, (p-1)a]$, lies in the $(p+1)$-th term of the lower central series of G (see for example Huppert [12], III.9.7) or, in other words, a group of exponent p satisfies the $(p-1)$-th Engel congruence.

If the answer is yes more generally, then any attempt to prove this will surely need to involve the derivation of suitable consequences of the exponent law — possibly in congruential form. Thus it has seemed worthwhile to seek relations and congruences which hold in groups of prime-power exponent. One way of finding congruences is to study groups of prime-power exponent which satisfy some additional condition. A possible additional condition to consider is the metabelian law. An advantage of this particular choice is that finitely generated metabelian groups of finite exponent are finite. Therefore there is a unique largest finite d-generator metabelian group of exponent p^k ; I will call it $M(d, p^k)$. The problem then is to get a good understanding of the relations which hold in $M(d, p^k)$. As tests of our understanding we can ask questions about the structure of $M(d, p^k)$. For example, following Burnside, what is the order of $M(d, p^k)$? or, more simply, what is the nilpotency class of $M(d, p^k)$?

2. Survey

The story begins with the case $k = 1$, which was studied by Meier-Wunderli in 1951 [18]. He obtained an excellent overview of the relations which hold in this case. Let $\{b_1, \ldots, b_d\}$ be a generating set for $M(d, p)$. Meier-Wunderli showed that the following relations hold (where $\emptyset$, for the empty word, denotes the identity element): for each d-tuplet $(m_1, \ldots, m_d)$ of integers such that $0 \leq m_i < p$ and $m_1 + \ldots + m_d = p$,

$$\prod_k [b_k, m_1b_1, \ldots, m_{k-1}b_{k-1}, (m_k-1)b_k, m_{k+1}b_{k+1}, \ldots, m_db_d]^{m_k} \neq \emptyset$$

where the product is taken over all k in $\{2, \ldots, d\}$ with $m_1 + \ldots + m_{k-1} = 0$ (and $m_k \neq 0$) .

From these it follows easily that $M(d, p)$ has nilpotency class at most p . Moreover, if $M(d, p) = M_1 \geq \ldots \geq M_{c+1} = [M_c, M] \geq \ldots \geq M_{p+1} = E$ is the lower central series of $M(d, p)$, then the elementary abelian section M_c/M_{c+1} has order p^r where

$$r = \begin{cases} (c-1)\binom{d+c-2}{c} & \text{for } 2 \leq c \leq p-1 , \\ (p-1)\binom{d+p-2}{p} - \binom{d+p-1}{p} + d & \text{for } c = p . \end{cases}$$

From this one can read off the order of $M(d, p)$ and that $M(d, p)$ has nilpotency class p for d at least 3 while the nilpotency class of $M(2, p)$ is $p-1$.

Results about the nilpotency class of $M(d, p^k)$ more generally were first obtained in the late sixties and there have been some more recent refinements. These results can be summed up as follows:

$$\text{cl } M(2, p^k) = k(p-1)p^{k-1} ,$$

$\geq$ Bachmuth, Heilbronn, and Mochizuki [2] ,
$\leq$ Dark and Newell [6] ;

$$\text{cl } M(p+1, p^k) \leq k(p-1)p^{k-1} + 1 ,$$

Bachmuth, Heilbronn, and Mochizuki [2] ;

$$\text{cl } M(d, p^k) = d(p^{k-1} - 1) + (p-2)p^{k-1} + 1$$

when $k \geq 2$ and $d \geq (p+2)(k-1)$,

$\leq$ Gupta, Newman, and Tobin [8] ,

$\geq$ Newman (unpublished, see appendix) ;

$$\text{cl } M(3, 4) = 5 ,$$

Gupta and Tobin [9] .

Some new results will be given later.

Gupta and Tobin [9] were also able to obtain the orders of $M(2, 4)$ and $M(3, 4)$; they are 2^{10} and 2^{34} . The only other published precise order determination is that $M(2, 8)$ has order 2^{63} (Hermanns [11]).

It is easy to get crude order bounds for $M(d, p^k)$. Let F be a free group of rank d and let $N = F^{p^k}F'$ be the subgroup of F generated by p^k-th powers and commutators. Since N has index p^{kd} in F, it has, by Schreier's formula, rank $1+(d-1)p^{kd}$. Clearly $M(d, p^k)$ is a quotient of $F/N^{p^k}N'$, so its order is a divisor of $p^{k+kd+k(d-1)p^{kd}}$. A very similar argument using $R = F^{p^{k-1}}F'$ and R^pR' gives that $p^{1+d(k-1)+(d-1)p^{(k-1)d}}$ divides $|M(d, p^k)|$. In particular $3^{12} \leq |M(2, 9)| \leq 3^{168}$. Had I done this calculation earlier it might have renewed my interest in metabelian groups of prime-power exponent sooner. As it was that had to wait till I saw a 1980 paper by Skopin [21] in which he proved that $|M(2, 9)|$ divides 3^{71} . Before I go into that, however, let me complete the survey.

It should be emphasised that these statements are merely convenient ways of summarising in comfortable and familiar terms results of detailed investigations of the groups. Let me close this survey with a statement which summarises some results in the, perhaps slightly less comfortable and less familiar, language of varieties of groups. Bryce [4] has shown that the variety of all metabelian groups of exponent p^k is the join of two subvarieties $\underline{\underline{U}}$ and $\underline{\underline{V}}$ where $\underline{\underline{U}}$ consists of all such groups whose Frattini subgroup has exponent p^{k-1} , or in other words — which are extensions of a group of exponent p^{k-1} by an elementary abelian group, and $\underline{\underline{V}}$ is a variety of nilpotent groups. In particular, for $p^k = 4$ the variety $\underline{\underline{U}}$ is the product variety $\underline{\underline{A}}_2\underline{\underline{A}}_2$.

3. Nilpotent quotient algorithm

Skopin's result suggested that the use of a computer implementation of the nilpotent quotient algorithm might shed new light on metabelian groups of prime-power exponent. A description of how the nilpotent quotient algorithm can be used to study groups satisfying an exponent law, or identical relation, is given in a paper by Havas and Newman [10]. Briefly the nilpotent quotient algorithm provides a consistent power-commutator presentation for a group of prime-power order; that is, a presentation on a set $\{a_1, \ldots, a_n\}$ of n generators with defining relations

$$a_i^p = \prod_{k=i+1}^{n} a_k^{\alpha(i, k)} ,$$

$$[a_j, a_i] = \prod_{k=j+1}^{n} a_k^{\alpha(i, j, k)} , \quad 1 \le i < j \le n ,$$

with consistency meaning the group has order p^n. In the Canberra implementation of the nilpotent quotient algorithm the presentation is built up class by class; here class is lower p-central class which is closely related to nilpotency class, it differs from the latter only in requiring the defining sections to have exponent p as well as be central. The crux of the building up process can be expressed in terms of the p-cover of a group. Let P be a d-generator group of prime-power order, then P is isomorphic to F/R where F is a free group of rank d. The group $F/R^p[R,F]$, which depends only on P, is the p-cover of P. Given a consistent power-commutator presentation for P it is easy to write down a power-commutator presentation for its p-cover. Then routine calculations reduce this to a consistent presentation. These calculations become lengthy once the presentations are at all large, and are best left to a machine.

From now on any power-commutator presentation mentioned will be consistent unless explicitly denied.

Given the size of groups handled in Havas and Newman [10] and given Skopin's upper bound for the order of $M(2, 9)$, it seems reasonable to expect that one can, with a machine, calculate a power-commutator presentation for $M(2, 9)$. The new problem is imposing the metabelian law. In principle this is simple; however, in practice one must, as always, produce a method which yields results at an acceptable cost. At the time Skopin's paper came to my attention (late in 1982) the Canberra implementation of the nilpotent quotient algorithm had had incorporated into it a subgroup handling capacity. (This is a refinement of a similar capacity designed by Oppelt in 1979 and described in his Aachen Diplomarbeit [20].) The program is called NQ 2.4 in Canberra and I will refer to it as such. In particular NQ 2.4 can calculate the commutator $[A, B]$ of two subgroups A and B of a group given by a power-commutator presentation and can calculate a power-commutator presentation for the quotient group to a central subgroup. Using this capacity it was possible to get a power-commutator presentation for $M(2, 9)$ in about an hour on the ANU VAX 11/780. This shows at once that the order of $M(2, 9)$ is 3^{57} and confirms the result of Bachmuth and Mochizuki [3] that the nilpotency class of $M(2, 9)$ is 12.*

Given the comparative ease with which this result was obtained, it is natural to

---

* At Groups-Korea 1983 Bachmuth informed me that he and Mochizuki had, during their work in the late sixties, determined that the order of $M(2, 9)$ is 3^{57}.

explore $M(3, 9)$ especially since the nilpotency class is at most 13 . Here the cost of enforcing the exponent 9 law (see Havas and Newman [10], p.218) soon becomes significant. In practice this difficulty is overcome by partially enforcing the exponent 9 law — that is, only requiring that a selected set of 9-th powers be trivial. Such a procedure is described briefly in section 3.4 of Havas and Newman [10]. The program NQ 2.4 applied to the 3-generator metabelian group M defined by the set of 9-th powers of all words of weight at most 6 yields a power-commutator presentation for the class 12 quotient in a few hours of VAX-time. From this power-commutator presentation one can read off the ranks of the first twelve lower central sections:

$$3 , 3 , 8 , 15 , 25 , 35 , 48 , 63 , 77 , 78 , 55 , 32 .$$

This suggests that the 13-th lower central section is small. Given that Bachmuth and Mochizuki [3] had said they had much evidence to indicate that the class of $M(3, 9)$ is 13 , there is strong motivation to go the one remaining class. Unfortunately it was clear that this next step would require a considerable amount of time with the main component being the time needed to calculate the second derived group. Had no better approach suggested itself I would have gone ahead. However it had become clear that the major reason why NQ 2.4 approaches the limits of its capacity on the VAX for this problem is that, because of its general nature, it does a lot of work which is not needed. Specifically it calculates for each relevant F/R , given by a power-commutator presentation, a power-commutator presentation for its p-cover $F/R^p[R, F]$; then, in effect, it calculates the second derived group of $F/R^p[R, F]$ and then the quotient group $(F/R^p[R, F])/(F/R^p[R, F])''$. However when F/R is metabelian this is simply $F/R^p[R, F]F''$. Thus if one could calculate a power-commutator presentation for $F/R^p[R, F]F''$ more directly there should be significant saving in time and also, because the presentations are smaller, some saving in space. However there is no other mechanism in NQ 2.4 to calculate such presentations. Without going into technicalities it is not possible to describe how one turns this observation into a suitable program. Suffice it to say that the data structure in NQ 2.4 is such that (fortunately for me) a relatively minor modification of the part which yields a not necessarily consistent power-commutator presentation for the p-cover allows one to calculate such a presentation for $F/R^p[R, F]F''$. With the modified program, NQM, about an hour of VAX-time gives the unexpected result that M , and therefore $M(3, 9)$, has nilpotency class 12 . The order of M is 3^{448} and so the order of $M(3, 9)$ divides 3^{448} . Every relation of M is a relation of $M(3, 9)$ so, for example, if $\{a, b, c\}$ generates $M(3, 9)$, then

$$[c, a, 10b] = [c, 3a, 8b] = [c, 5a, 6b] = [c, 7a, 4b] = [c, 9a, 2b] .$$

Until one has a power-commutator presentation for $M(3, 9)$ itself there is no guarantee that some useful consequence of the exponent law has not been overlooked.

The set of 9-th power words used in NQ 2.4 (and hence also in NQM) for testing whether exponent 9 holds is too large for comfort. However in metabelian groups it is possible to get a much smaller test set of 9-th powers using the following simple result.

Let G *be a metabelian group and* e *a positive integer. If* $g^e = u^e = v^e = (gu)^e = (gv)^e = \emptyset$ *for* g *in* G *and* u, v *in* G', *then*

(a) $(guv)^e = \emptyset$, *and*

(b) *if* G' *has exponent* e, *then* $(g[u, x])^e = \emptyset$ *for all* x *in* G.

Proof. (a) Because G is metabelian standard collection yields

$$(gu)^n = g^n u^n \prod_{i=1}^{n-1} [u, ig]^{\binom{n}{i+1}}$$

for every positive integer n. Hence

$$\prod_{i=1}^{e-1} [u, ig]^{\binom{e}{i+1}} = \emptyset$$

and

$$\prod_{i=1}^{e-1} [v, ig]^{\binom{e}{i+1}} = \emptyset .$$

Again using that G is metabelian yields

$$\prod_{i=1}^{e-1} [uv, ig]^{\binom{e}{i+1}} = \emptyset$$

and the result follows easily.

(b) Note first that (a) yields $(gu^r)^e = \emptyset$ for all r and so, in particular for $r = -1$. Also (a) gives

$$(g[g, x^{-1}]u)^{ex} = \emptyset$$

because $[g, x^{-1}]^e = \emptyset$ and $(g[g, x^{-1}])^e = g^{x^{-1}e} = \emptyset$. Hence $(gu^x)^e = \emptyset$ and therefore

$$(g[u, x])^e = (gu^{-1}u^x)^e = \emptyset$$

by (a).

From this one can deduce that a 3-generator metabelian group has exponent 9 provided a suitable set of a few hundred 9-th powers is trivial. Testing such a set relative to the power-commutator presentation for M yields that M has exponent 9 in less than 20 minutes on the VAX. Hence M is $M(3, 9)$; so we have a power-commutator presentation for $M(3, 9)$ and $M(3, 9)$ has order 3^{448}.

The power-commutator presentation provides very detailed information about $M(3, 9)$. One can read off relations, such as those given earlier, and much other structural detail. For example, the commutator subgroup $M(3, 9)'$ is the direct sum of 76 cyclic subgroups of order 9 and 290 cyclic subgroups of order 3 ; the quotient group $M(3, 9)/(M(3, 9)')^3$ has nilpotency class 12 — this contrasts with: $M(2, 9)/(M(2, 9)')^3$ has nilpotency class 10 . It needs to be emphasised that this is quite a big group; so that, while there is much information in the presentation, extracting it often requires effort. In this case, using such usual conventions as omitting all trivial commutators, the printed version of the power-commutator presentation runs to some 50 pages. It is somewhat tedious and error-prone to use this for systematic searching by hand. It is usually more convenient and accurate to rely on a machine; NQ 2.4 (and its derivative NQM) has the capacity to allow for such searching.

It should further be remarked that there is information obtained during the running of these programs which sheds light on the structure of the group being investigated other than that which appears in the power-commutator presentation. For example $M(3, 9)$ can be defined as a metabelian group by 88 9-th powers and no fewer; moreover the program gives a set of 9-th powers which suffice for this.

4. Exponent four

As part of its acceptance trials NQM was used to obtain power-commutator presentations for $M(2, 4)$, $M(3, 4)$ and $M(2, 8)$. From these one can confirm the orders obtained by Gupta and Tobin [9] and by Hermanns [11], and also the more detailed results of these papers. The current version of NQM takes less than 5 seconds on the VAX to calculate the power commutator presentations for $M(2, 4)$ and $M(3, 4)$. Moreover it yields that $M(2, 4)$ can be defined as a metabelian group by 6 fourth powers and $M(3, 4)$ by 22 .

It is irresistible to push on. Using just the old exponent enforcing procedure yields a power-commutator presentation for $M(4, 4)$ in less than half a minute and one for $M(5, 4)$ in about 8 minutes. From these one can confirm that the nilpotency classes are 5 and 6 respectively and read off that their orders are 2^{93} and 2^{224}. Moreover 59 and 130 fourth powers suffice and are needed to define them as metabelian groups.

Let $M(X, 4)$ be the free metabelian group of exponent 4 freely generated by a set X . Using detailed information from the above power-commutator presentations and results of Kovács and Newman [16] about the free groups of the product variety $\underline{\underline{A}}_2\underline{\underline{A}}_2$, one can determine a basis for the commutator subgroup $M(X, 4)'$. Let $<$ be a linear order on X , then $M(X, 4)'$ is the direct sum of cyclic subgroups of order 4 generated by $\{[x, y] : x < y\}$ and an elementary abelian group with basis the union of the following sets (where all variables are universally quantified): for $c \geq 3$,

$$\{[y, x_1, \ldots, x_{c-1}] : x_1 < \ldots < x_{c-1}, x_1 < y_1, y \notin \{x_2, \ldots, x_{c-1}\}\} ,$$

$$\{[x_1, y, y, x_2, \ldots, x_{c-2}] : x_1 < \ldots < x_{c-2}, y \notin \{x_1, \ldots, x_{c-2}\}\},$$

$$\{[x, y, z, w, w] : x < y, z < w, y \neq z\},$$

$$\{[x, y, y, z, z] : x \neq y < z \neq x\},$$

$$\{[x, y, z, z] : x < y, z\}.$$

That these elements generate $M(X, 4)'$ can be proved using the relations in Gupta and Tobin [9] and the relation

$$[z, x, y, w, w][w, x, y, z, z][w, x, z, y, y][w, y, z, x, x] = \emptyset$$

which is a straight-forward consequence of $(xyzw)^4 = \emptyset$. The independence of the elements follows from the power-commutator presentation for $M(4, 4)$ and result 4.05 of Kovács and Newman [16]. A simple corollary is that for every positive integer d the order of $M(d, 4)$ is

$$2^{1+d+(d-1)2^d+3\binom{d}{2}+5\binom{d}{3}+2\binom{d}{4}}.$$

One can from the working of NQM deduce a minimal set of fourth powers defining $M(X, 4)$ as a metabelian group, namely, the union of the following:

$$\{x^4\}, \quad \{(xy)^4, (xy^{-1})^4 : x < y\}, \{(xy^2)^4 : x \neq y\},$$

$$\{(xyz)^4 : x < y < z\}, \{(xyz^2)^4 : x < y\},$$

$$\{(xy[x, z])^4 : x < y\},$$

and $$\{(xyzw)^4, (xzwy)^4, (xwyz)^4 : x < y < z < w\}.$$

Hence $M(d, 4)$ can be defined as a metabelian group by $d + 4\binom{d}{2} + 7\binom{d}{3} + 3\binom{d}{4}$ fourth powers.

Furthermore one can sharpen the result of Bryce to: the variety $\underline{\underline{M}}_4$ of metabelian groups of exponent 4 is the join of $\underline{\underline{A}}_2\underline{\underline{A}}_2$ and a variety of nilpotent groups of class at most 5. Indeed analysis of the subvarieties of $\underline{\underline{M}}_4$ can be carried to a satisfactory conclusion. Let me recall that the lattice of subvarieties of $\underline{\underline{A}}_2\underline{\underline{A}}_2$ is a single chain (see p.133 of Kovács and Newman [16])

$$\underline{\underline{E}}, \underline{\underline{A}}_2, \underline{\underline{A}}_4 = \underline{\underline{I}}_1, \underline{\underline{I}}_2, \underline{\underline{I}}_3, \underline{\underline{I}}_{4*}, \underline{\underline{I}}_4, \ldots, \underline{\underline{I}}_\omega = \underline{\underline{A}}_2\underline{\underline{A}}_2$$

where $\underline{\underline{I}}_c$ is the subvariety of groups of nilpotency class at most c and $\underline{\underline{I}}_{2k*}$ $(k \geq 2)$ is defined by the law

$$\prod_{i=2}^{2k} [x_i, x_1, \ldots, x_{i-1}, x_{i+1}, \ldots, x_{2k}].$$

These varieties are clearly join-irreducible. There are exactly three further join-irreducible varieties in $\underline{\underline{M}}_4$. They are

(i) $\underline{\underline{I}}_\alpha$ which consists of all groups of class at most 4 which satisfy the law $[x, y, z, w][x, z, y, w][x, w, y, z]$;

(ii) $\underline{\underline{I}}_\beta$ which consists of all groups of class at most 4 which satisfy the law $[x, y]^2$;

(iii) $\underline{\underline{I}}_\gamma$ which consists of all groups of class at most 5 which satisfy the law $[x, y]^2$.

Clearly $\underline{\underline{I}}_\beta < \underline{\underline{I}}_\gamma$, and $\underline{\underline{I}}_\alpha \not\leq \underline{\underline{I}}_\gamma$ because the quotient of $M(2, 4)$ which satisfies the Engel law $[x, y, y, y]$ is not in $\underline{\underline{I}}_\gamma$. Moreover it is straight-forward to check that the following meet relations hold:

$$\underline{\underline{I}}_\alpha \wedge \underline{\underline{A}}_2\underline{\underline{A}}_2 = \underline{\underline{I}}_{4*}, \quad \underline{\underline{I}}_\beta \wedge \underline{\underline{A}}_2\underline{\underline{A}}_2 = \underline{\underline{I}}_4, \quad \underline{\underline{I}}_\gamma \wedge \underline{\underline{A}}_2\underline{\underline{A}}_2 = \underline{\underline{I}}_5 .$$

The lattice of subvarieties of $\underline{\underline{M}}_4$ can be shown to be distributive. Hence, because every element of a distributive lattice can be uniquely written as a join of join-irreducible elements, one has a complete description of the lattice of subvarieties of $\underline{\underline{M}}_4$. It can be pictured as follows:

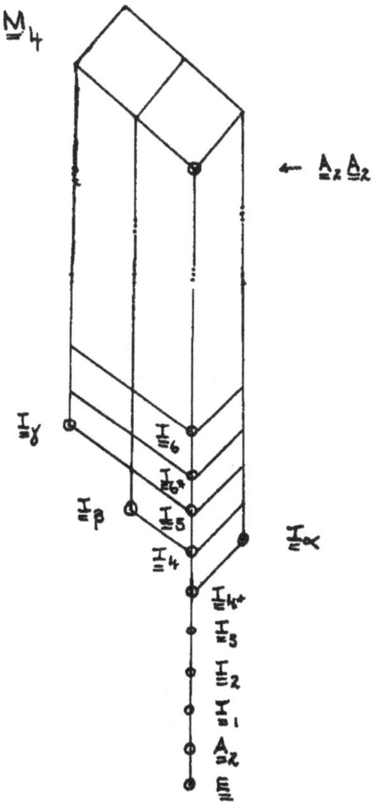

As a further strengthening of Bryce's result one gets

$$\underline{\underline{M}}_4 = \underline{\underline{A}}_2\underline{\underline{A}}_2 \vee \underline{\underline{I}}_\alpha \vee \underline{\underline{I}}_\gamma \, .$$

5. Some further results

So far I have mentioned some of the results which can be obtained from relatively short runs of the program NQM. Let me conclude by giving a brief indication of how much further it has been possible to go with such calculations. For exponent 9 it has been possible to get a power-commutator presentation for the 4-generator free (metabelian) group. It has class 13 and order 3^{2338}. Similar calculations have been done for exponent 8 yielding that $|M(4, 8)| = 2^{2232}$ and confirming that cl $M(4, 8)$ is 13 (though only just, for the last non-trivial term of the lower central series has order 2^4). On the way to this result one gets that $|M(3, 8)| = 2^{450}$ and, again somewhat unexpectedly, that cl $M(3, 8)$ is 12. (In a recent letter Newell states he has obtained this class result independently.) It has been confirmed that $M(2, 16)$ has class 32; it has order dividing 2^{376}. These statements are in each case the tip of an underlying power-commutator presentation. These presentations obviously contain lots of information. However they are so big, for example that for $M(4, 9)$ would be over 500 pages long, that they are nearly unmanageable except in a machine. But with a machine one can extract information readily. As an example, Newell asked for confirmation that the fourth power $(M(4, 8)_5)^4$ of the fifth term of the lower central series of $M(4, 8)$ is not the identity. In fact, one finds that $[a_1, a_2, a_3, a_4, a_1]^4 = \emptyset$ from which it can be deduced that $[a_1, a_2, a_3, a_4, a_5]^4 = \emptyset$ in all metabelian groups of exponent 8. On the other hand $[a_1, a_2, a_3, a_4]^4$ is non-trivial.

Appendix

This appendix reproduces in minimally edited form an aide-memoire dated February 1969. At about the same time Gupta [7] constructed by quite different methods an example for the case $k = 2$.

Theorem 3 of Gupta, Newman and Tobin [8] is best possible in the following sense.

For $n \geq 2$ there is an n-generator metabelian group of exponent p^k with nilpotency class at least

$$(n-1)(p^{k-1}-1) + p^{k-1}(p-1) \, .$$

Let A be a cyclic group of order p generated by a, let U be a cyclic group of order p^k generated by u, and let V be the direct product of $n-1$ cyclic groups of order p^{k-1} generated by $\{v_1, \ldots, v_{n-1}\}$. Let $W = A \text{ wr } (U * V)$, let B be the base group (*i.e.*, the functions from $U \times V$ to A), and N the subgroup of B consisting of those functions f for which

$$\prod_{i=0}^{p-1} f(u^{j+ip^{k-1}}v) = e \quad \text{(the identity)}$$

for all $j \in \{0, \ldots, p^{k-1}-1\}$ and all v in V. Clearly N admits the action of $U \times V$ on B. Since W is metabelian, so is $(U \times V)N$. Moreover $(U \times V)N$ has exponent p^k because for t in $U \times V$ and f in N

$$(tf)^{p^k} = (u^{ip^{k-1}}f^*)^p$$

for some $i \in \{0, \ldots, p-1\}$ and $f^* \in N$, and so $(tf)^{p^k} = e$ from the definition of N. Let $\bar{a} \in B$ be defined by $\bar{a}(e) = a$, $\bar{a}(t) = e$ for $t \neq e$. A routine induction establishes that, for $r \in \{0, \ldots, p^k-1\}$,

$$[\bar{a}, ru](u^j) = a^{(-1)^{r-j}\binom{r}{j}} \quad \text{for } 0 \le j \le r,$$

$$[\bar{a}, ru](t) = e, \quad t \notin \{u^0, \ldots, u^r\}.$$

Hence $[\bar{a}, p^{k-1}u] \in N$. The claim is established by showing that the subgroup of $(U \times V)N$ generated by $\{u[\bar{a}, p^{k-1}u], v_1, \ldots, v_{n-1}\}$ has the required nilpotency class. It suffices to show

$$[u[\bar{a}, p^{k-1}u], (p^{k-1}-1)v_1, \ldots, (p^{k-1}-1)v_{n-1}, (p^{k-1}(p-1)-1)u[\bar{a}, p^{k-1}u]]$$

is non-trivial. But after a little manipulation this commutator becomes $[\bar{a}, (p^k-1)u, (p^{k-1}-1)v_1, \ldots, (p^{k-1}-1)v_{n-1}]$ which is non-trivial (see, for example, Liebeck [17], pp.448-449).

References

1 W.A. Alford and Bodo Pietsch, "An application of the nilpotent quotient algorithm", *Burnside Groups* (Proc. Workshop, Univ. Bielefeld, Bielefeld, 1977), pp. 47-48 (Lecture Notes in Mathematics, **806**. Springer, 1980), MR82d:20001.

[2] S. Bachmuth, H.A. Heilbronn, and H.Y. Mochizuki, "Burnside metabelian groups", *Proc. Roy. Soc. London Ser. A* **307** (1968), pp. 235-250, MR38:4561.

[3] S. Bachmuth and H.Y. Mochizuki, "The class of the free metabelian group with exponent p^2", *Comm. Pure Appl. Math.* **21** (1968), pp. 385-399, MR38:234.

[4] R.A. Bryce, "Varieties of metabelian p-groups", *J. London Math. Soc.* (2) **13** (1976), pp. 363-380, MR54:12896.

[5] W. Burnside, "On an unsettled question in the theory of discontinuous groups", *Quart. J. Pure Appl. Math.* **33** (1902), pp. 230-238, FdM33,149.

[6] Rex S. Dark and Martin L. Newell, "On 2-generator metabelian groups of prime-power exponent", *Arch. der Math.* **37** (1982), pp. 385-400, MR83e:20031.

[7] N.D. Gupta, "The free metabelian group of exponent p^2", *Proc. Amer. Math. Soc.* **22** (1969), pp. 375-376, MR39: 6984.

[8] N.D. Gupta, M.F. Newman, and S.J. Tobin, "On metabelian groups of prime-power exponent", *Proc. Roy. Soc. London Ser. A* **302** (1968), pp. 237-242, MR36:5222.

[9] Narain D. Gupta, and Seán J. Tobin, "On certain groups with exponent four", *Math. Z.* 102 (1967), pp. 216-226, MR36:5220.

[10] George Havas and M.F. Newman, "Application of computers to questions like those of Burnside", *Burnside Groups* (Proc. Workshop. Univ. Bielefeld, Bielefeld 1977), pp. 211-230 (Lecture Notes in Mathematics, 806. Springer Verlag, Berlin, Heidelberg, New York, 1980), MR 82d:20002.

[11] Franz-Josef Hermanns, "Eine metabelsche Gruppe vom Exponenten 8", *Arch. der Math.* 29 (1977), pp. 375-382, MR57:12697.

[12] B. Huppert, *Endliche Gruppen I* (Die Grundlehren der Mathematischen Wissenschaften, 134. Springer-Verlag, Berlin, Heidelberg, New York, 1967), MR37:302.

[13] A.I. Kostrikin, "On Burnside's problem", *Dokl. Akad. Nauk SSSR (N.S.)* 119 (1958), pp. 1081-1084, MR24:A1948.

[14] A.I. Kostrikin, "The Burnside problem", *Izv. Akad. Nauk SSSR Ser. Mat.* 23 (1959), pp. 3-34; *Amer. Math. Soc. Transl.* (2) 36 (1964), pp. 63-99, MR24:A1947.

[15] A.I. Kostrikin, "Sandwiches in Lie algebras", *Mat. Sb. (N.S.)* 110 (152) (1979), pp. 3-12; *Math. USSR Sb.* 38 (1981), pp. 1-9, MR83e:17004.

[16] L.G. Kovács and M.F. Newman, "On non-Cross varieties of groups", *J. Austral. Math.* 12 (1971), pp. 129-144, MR45:1966.

[17] Hans Liebeck, "Concerning nilpotent wreath products", *Proc. Cambridge Philos. Soc.* 58 (1962), pp. 443-451, MR25:3087.

[18] H. Meier-Wunderli, "Metabelsche Gruppen", *Comment Math. Helv.* 25 (1951), pp. 1-10, MR12,671.

[19] M.F. Newman, "Problems", *Burnside Groups* (Proc. Workshop, Univ. Bielefeld, Bielefeld 1977), pp. 249-254 (Lecture Notes in Mathematics, 806. Springer-Verlag, Berlin, Heidelberg, New York, 1980), MR82a:20048.

[20] Elmar Oppelt, "SOGOS Ein Programmsystem zur Handhabung von Untergruppen aufloesbarer Gruppen" (Diplomarbeit, RWTH, Aachen, 1981).

[21] A.I. Skopin, "Metabelian groups of exponent 9 with two generators", *Zap. Nauchn. Sem. Leningrad. Otdel. Mat. Inst. Steklov. (LOMI)* 103 (1980), pp. 124-131, 159-160, MR83d:20022.

Department of Mathematics,
Institute of Advanced Studies,
The Australian National University,
Canberra, A.C.T. 2601
Australia

PROCEEDINGS OF 'GROUPS — KOREA 1983'
KYOUNGJU, August 1983

05C25, 20B15,
20D08

SYMMETRIC GRAPHS AND THE CLASSIFICATION OF THE FINITE SIMPLE GROUPS

Cheryl E. Praeger

Abstract

Some applications of the finite simple group classification to the study of symmetric graphs are discussed.

1. Introduction

Since the classification of the finite simple groups was completed in 1980 several longstanding problems in the theory of finite permutation groups have been solved: that is solutions have been found which rely on the simple group classification, see for example [7], [8], [18]. It is widely believed that similar techniques should produce solutions for problems about symmetric graphs, and although some problems have been solved there are not as many positive results as one might expect. In this paper I will discuss the effects of the simple group classification on several problems about symmetric graphs.

A simple undirected graph Γ with vertex set V and edge set E is called G-*symmetric*, where G is a group of automorphisms of Γ, if G is transitive on V and on the set $E' = \{(x, y) | \{x, y\} \in E\}$ of directed edges of Γ. Equivalently Γ is G-symmetric if G is transitive on V and the stabilizer G_x of a vertex x in G is transitive on the set $\Gamma_1(x) = \{y | \{x, y\} \in E\}$ of neighbours of x. We shall usually assume that Γ is connected. We shall say that Γ is *symmetric* if it is G-symmetric for some $G \leq \mathrm{Aut}\,\Gamma$.

Symmetric graphs arise naturally in the theory of finite permutation groups as follows. If G is a transitive permutation group on a finite set V then a natural action of G on the set $V \times V$ is induced by defining $g : (x, y) \to (x^g, y^g)$ for $g \in G$, $x, y \in V$. For each nondiagonal orbit U of G in $V \times V$ (that is $U \neq \{(x, x) | x \in V\}$), we define the directed graph $\Gamma(U)$ to be the graph with vertex set V and (directed) edge set U. If for some $(x, y) \in U$ we also have $(y, x) \in U$ then this property is true for all elements (x, y) in U: we say that U is *self paired* and in this case $\Gamma(U)$ can be regarded as an undirected graph. Indeed $\Gamma(U)$ is G-symmetric by definition. Moreover G has such a self paired orbit U if and only if the order of G is even ([24], 16.5). Finally the (directed) graphs may or may not be connected, but we know that $\Gamma(U)$ is connected for all nondiagonal U if and only if G is primitive on V ([14], 1.12).

This link between symmetric graphs and transitive permutation groups should give us some clues for discovering applications of the classification of simple groups. For permutation groups, the most spectacular consequences of the classification have been in problems about primitive permutation groups. However if Γ is a G-symmetric graph the primitivity of the action of G on vertices has rarely played an important role in studying Γ except in very special cases, for example if Γ is distance transitive. Thus there have not been as many new results about symmetric graphs using the classification as perhaps one might have hoped. The results which have been proved up to the present, using the classification, are about G-symmetric graphs for which G is transitive on paths of length 2. We shall discuss these results and several other problems in sections 2-4. In section 5 we investigate a new approach to these problems in the case where G is primitive on vertices.

2. s-transitive graphs

An *s-path* in a graph Γ, s a positive integer, is an $(s+1)$-tuple $(x_0, x_1, \dots, x_s)$ of vertices with $x_i \in \Gamma_1(x_{i-1})$ if $1 \le i \le s$ and $x_i \ne x_{i-2}$ if $2 \le i \le s$. A symmetric graph Γ is called *s-transitive* if $G = \mathrm{Aut}\,\Gamma$ is transitive on the set of all s-paths in Γ. Symmetric graphs are 1-transitive by definition, and many of them are s-transitive for larger values of s. Interest in s-transitive graphs stems from work of Tutte [21], [22] who showed that if Γ is s-transitive for some $s \ge 1$ and Γ has valency 3 then $s \le 5$. Further if Γ contains a circuit and $G = \mathrm{Aut}\,\Gamma$ is transitive on the set of all 2-paths then it can easily be shown that the stabilizer G_x of a vertex x is 2-transitive on $\Gamma_1(x)$. Since Tutte's result was proved most of the work on s-transitive graphs has involved the imposition of extra conditions on the 2-transitive constituent $G_x^{\Gamma_1(x)}$. Two surveys of s-transitive graphs up to 1979 can be found in [13] and [23]. The results are mainly due to Gardiner and Weiss and are directed at two problems:

PROBLEM 2.1. *Show that if* Γ *is* s*-transitive of valency at least* 3 *then* s *is bounded (perhaps* $s \ge 7$ *). Classify those* s*-transitive graphs where* s *is large (perhaps* $s \ge 4$ *).*

PROBLEM 2.2. *Show that if* Γ *is* 2*-transitive then the order of the subgroup of* $G = \mathrm{Aut}\,\Gamma$ *fixing* $\Gamma(x) \cup \Gamma(y)$ *pointwise, where* $\{x, y\}$ *is an edge, is bounded by a function of the valency of* Γ.

(We note that simple circuits are s-transitive for all s, hence the restriction to a valency $v \ge 3$.) Problem 2.2 is a special case of the following more general problem which is the analogue for symmetric graphs of the famous "Sims' conjecture" for primitive permutation groups (see [11]).

PROBLEM 2.3. *Show that if* Γ *is* G*-symmetric of valency* v *such that* G_x *is primitive on* $\Gamma_1(x)$*, where* x *is a vertex, then* $|G_x|$ *is bounded by a function of* v.

Sims' conjecture for primitive permutation groups has been proved (see [11]), using the simple group classification but Problem 2.3 is still open. The following theorem, formulated by Gardiner [12], (2.3), for undirected graphs, and originally for primitive groups by Thompson [20], reduces Problem 2.3 to bounding the order of a subnormal subgroup of G_x of prime power order.

THEOREM 2.1. *Let* Γ *and* G *be as in Problem 2.3. Then for each edge* $\{x, y\}$ *the pointwise stabilizer in* G *of* $\Gamma_1(x) \cup \Gamma_1(y)$ *is a* p*-group for some prime* p.

The next theorem summarizes the status of Problems 2.1 and 2.2 before the classification (see [13], p.28); the results are due mainly to Gardiner and Weiss.

THEOREM 2.2. *Let* Γ *be* s*-transitive of valency* $v \geq 3$ *with* $s \geq 2$. *Set* $G = \mathrm{Aut}\,\Gamma$ *and let* $G_1(x, y)$ *be the pointwise stabilizer of* $\Gamma_1(x) \cup \Gamma_1(y)$ *where* $\{x, y\}$ *is an edge. Then* <u>*one*</u> *of the following is true:*

(a) $G_1(x, y) = 1$;

(b) $s \leq 3$;

(c) $4 \leq s \leq 7$ *and* $|G_1(x, y)|$ *is bounded* ;

(d) $G^{\Gamma_1(x)}$ *is a* 2*-transitive group "not known in 1979".*

One of the major consequences of the classification of the finite simple groups is the classification of all finite 2-transitive permutation groups (see [7], Theorem 5.3(S)); there are no 2-transitive groups which were not known in 1979. Using the simple group classification which (following Cameron), I shall denote by (S), Weiss was able to show that $s \leq 7$ in Problem 2.1 (see [7], p.20, and [23], p.843). His result, together with some information from [9], Theorem 2, follows.

THEOREM 2.3 (S). *Let* Γ *be* s*-transitive of valency* $v \geq 3$ *where* $s \geq 2$, *let* $G = \mathrm{Aut}\,\Gamma$ *and let* $G_1(x, y)$ *be the pointwise stabilizer of* $\Gamma_1(x) \cup \Gamma_1(y)$ *where* $\{x, y\}$ *is an edge. Then the following are true:*

(a) $s \leq 7$, *and*

(b) *there is a function* $f(v)$ *such that either*

(i) $|G_1(x, y)| \leq f(v)$

or (ii) $\mathrm{PSL}(n, q) \leq G_x^{\Gamma_1(x)} \leq \mathrm{P\Gamma L}(n, q)$

in its natural representation with $n \geq 3$ *and* q *a power of the prime* p *of Theorem 2.1, and either* $s = 2$, *or* $s = 3$ *and* p *is* 2 *or* 3. *Moreover* Γ *has girth at least* 6.

Thus Problem 2.2 is also nearly solved. The general Problem 2.3 of bounding the order of a vertex stabilizer is still open and is probably very difficult. Of course the solution follows from the proof of Sims' Conjecture for G-symmetric graphs where G is primitive on vertices, and it is not too difficult to extend this a little. Define

$$G^{+} = \langle G_x \mid x \in V \rangle .$$

LEMMA 2.4. *For a G-symmetric graph Γ, either*

(a) $G^{+} = G$ *(and Γ is not bipartite), or*

(b) $|G : G^{+}| = 2$, *Γ is bipartite,*

and the two orbits of G^{+} in V are the blocks of a bipartition.

Proof. Since G^{+} contains G_x, G^{+} has $|G : G^{+}| = a$ orbits in V, and if Γ is bipartite then G^{+} preserves any bipartition so that $a > 1$. Suppose then that $a > 1$. Let A be an orbit of G^{+}, let $x \in A$, $y \in \Gamma_1(x)$, and let B be the G^{+}-orbit containing y. Then $B \neq A$ as Γ is connected and $a > 1$, and $\Gamma_1(x) \subseteq B$ as G_x is transitive on $\Gamma_1(x)$. Similarly $\Gamma_1(y) \subseteq A$, and as Γ is connected $V = A \cup B$ and Γ is bipartite.

The graph Γ is called *G-semiprimitive* if G^{+} is primitive on each of its orbits in V. We can show that Problem 2.3 is solved for G-semiprimitive graphs.

THEOREM 2.5 (S). *Suppose that Γ is G-symmetric and G-semiprimitive of valency $v \geq 3$ such that G_x is primitive on $\Gamma_1(x)$, x a vertex. Then $|G_x|$ is bounded by a function of v.*

Proof. If G is primitive on V then by [11] there is an increasing integral function f such that $|G_x| \leq f(v)$. Suppose therefore, by Lemma 2.4 that Γ is bipartite, G^{+} has two orbits A, B in V and is primitive on each. Let K be the kernel of the action of G^{+} on A. Suppose first that $K = 1$. Let $x \in A$ and let Δ be an orbit of G_x in A contained in the set $\Gamma_2(x)$ of vertices at distance 2 from x. Then $|\Delta| \leq |\Gamma_2(x)| \leq v(v-1)$, and, by [11], $|G_x| = |G_x^A| \leq f(|\Delta|) \leq f(v(v-1))$. Suppose then that K is nontrivial. Then as G^{+} is primitive on B, K is transitive on B. However as K fixes $x \in A$, K fixes $\Gamma_1(x) \subseteq B$ setwise. Thus $\Gamma_1(x) = B$, $\Gamma = K_{v,v}$ is the complete bipartite graph, and $|G_x| \leq v!(v-1)!$. This completes the proof.

To obtain a complete solution for Problem 2.3 it may be helpful to understand the nature of blocks of imprimitivity on vertices of G-symmetric graphs Γ which are not semiprimitive and for which the vertex stabilizers G_x are primitive on their neighbours $\Gamma_1(x)$. At present this is not well understood and may well be a necessary step to a solution.

3. Circuits in 2-transitive graphs

In this section we assume that Γ is a connected 2-transitive graph of valency $v \geq 3$ and that $G \leq \mathrm{Aut}\,\Gamma$ is transitive on vertices and 2-paths. We recall that in this situation the stabilizer G_x is 2-transitive on $\Gamma_1(x)$. The girth of Γ is 3 if and only if Γ is the complete graph on $v+1$ vertices; (the girth is the length of a shortest circuit in Γ). So we shall assume that Γ has girth at least 4 . Suppose first that Γ has girth 4 . Then as G is transitive on 2-paths it follows that for vertices x, y at distance 2 in Γ the number $k = |\Gamma_1(x) \cap \Gamma_1(y)| > 1$ of 2-paths from x to y is independent of the choice of x and y ; the parameter k gives a measure of the number of circuits of length 4 in Γ . The analogous situation to this for primitive permutation groups was considered by Manning in 1927 and 1929. He showed that $k < v-1$ and that if G_x is 2-primitive on $\Gamma_1(x)$ then G_x is faithful on $\Gamma_1(x)$ ([24], 17.7). His results were generalized by Cameron [4], [5] and give very good results for the present problem about 2-transitive graphs. His results, discussed in [5] include:

(a) if $k > v/2$ then Γ is the incidence graph of a symmetric design ([5], Theorem 4.1), and

(b) if G_x acts on $\Gamma_1(x)$ as A_v or S_v then Γ is one of a known list of graphs ([5], Theorem 4.5).

Cameron conjectured that if G is primitive on V then the parameter k can be at most 6: and indeed using the simple group classification the conjecture has been nearly, but not completely proved, namely the case where $G_x^{\Gamma_1(x)}$ has a nontrivial abelian normal subgroup has not been settled (see [9] and [10]).

THEOREM 3.1 (S). *Let Γ be a connected graph of valency $v \geq 3$ such that $G \leq \mathrm{Aut}\,\Gamma$ is transitive on vertices and for a vertex x, $G_x^{\Gamma_1(x)}$ is 2-transitive with no nontrivial abelian normal subgroup. If two vertices at distance 2 are joined by more than six paths of length 2 then either Γ is the incidence graph of a known symmetric design or Γ is a dual orthogonal graph.*

PROBLEM 3.1. *Determine the 2-transitive graphs Γ for which the constituent $G_x^{\Gamma_1(x)}$ has a nontrivial abelian normal subgroup and $k > 6$.*

Perhaps also we could suggest the more difficult problem:

PROBLEM 3.2. *Determine the 2-transitive graphs Γ with $k > 2$.*

Now if the girth is greater than 4 it is difficult to say anything in general about the case where G_x is faithful on $\Gamma_1(x)$. However when G_x is not faithful on $\Gamma_1(x)$ it turns out that in many cases each pair of vertices at distance 3 is joined by the same number of paths of length 3. We have the following information (which is not entirely satisfactory) about 2-transitive graphs in which G_x is unfaithful on $\Gamma_1(x)$ (see [9], [17], [23]).

THEOREM 3.2 (S). *Let* Γ *be a connected graph of valency* $v \geq 3$ *such that* $G \leq \mathrm{Aut}\,\Gamma$ *is transitive on vertices and for a vertex* x , G_x *is 2-transitive and unfaithful on* $\Gamma_1(x)$ *and this constituent has no nontrivial abelian normal subgroup. Then*

(a) *if* Γ *has girth* 4 , *then* Γ *is the complete bipartite graph* $K_{v,v}$, *the incidence graph of points and hyperplanes of a projective geometry* $\mathrm{PG}(n, q)$, $n \geq 3$, *or a dual orthogonal graph,*

(b) Γ *does not have girth* 5 ,

(c) *if* Γ *has girth* 6 *and if* $G_x^{\Gamma_1(x)}$ *is not a normal extension of* $\mathrm{PSL}(n, q)$, $n \geq 3$, *then* G *is transitive on* 3*-paths and every pair of vertices at distance* 3 *is joined by a constant number* k *of paths of length* 3. *Moreover if* $k > 6$ *then either* Γ *is the incidence graph of a projective plane* (*and* $k = v$), *or* $G_x^{\Gamma_1(x)}$ *is a normal extension of a Suzuki group* $\mathrm{Sz}(q)$ (*and* $k = q$), *a unitary group* $\mathrm{PSU}(3, q)$ (*and* $k = q$ *or* q *is* 3 *or* 5), *or the smallest Ree group* $R(3)$.

It is necessary to make a few remarks about this result. It is basically Theorem 2 of [17]. However in the case where $G_x^{\Gamma_1(x)}$ is a normal extension of $\mathrm{PSU}(3, q)$ it follows from [23] Theorem 4.8 that the subgroup K of G_x fixing $\Gamma_1(x)$ pointwise acts faithfully on $\Gamma_1(y)$ for $y \in \Gamma_1(x)$. Then by [16] Lemma 1.9 and Theorem 2 it follows that G is transitive on 3-paths and two vertices at distance 3 are joined by a constant number k of paths of length 3. Finally the argument in [17], p.137 shows that k is 1, 2 or q , or q is 3 or 5.

Again as in the previous section it is the projective groups $\mathrm{PSL}(n, q)$, $n \geq 3$ which cause problems with the general result. One might ask:

PROBLEM 3.3. *Let* Γ *be* G*-symmetric of girth at least* 6 *and for* $x \in V$ *assume that* $\mathrm{PSL}(n, q) \leq G_x^{\Gamma_1(x)} \leq \mathrm{P\Gamma L}(n, q)$ *in its natural representation,* $n \geq 3$.

(a) *For* $y \in \Gamma_1(x)$ *show that the subgroup* $G_1(x, y)$ *fixing* $\Gamma_1(x) \cap \Gamma_1(y)$ *has order bounded by a function of* n *and* q (*see Problem 2.2*).

(b) *If* G_x *is not faithful on* $\Gamma_1(x)$ *determine whether or not* G *is transitive on* 3*-paths, and if so determine the possible values of* k . (*Note that by* [17] *Lemma* V, G *has at most two orbits on* 3*-paths, and in the transitive case* k *is* 1, 2, q *or* $q+1$.)

In section 5 we discuss 2-transitive graphs with a group of automorphisms primitive on vertices. We investigate a different approach to these problems, and show for example that in the situation of Theorem 3.2 (S), $S \leq G \leq \mathrm{Aut}\, S$ for some nonabelian simple group S (Corollary 5.4 (S)).

4. Distance transitive graphs

We could define a graph Γ to be (G, s)*-symmetric* if G is transitive on the sets $\Gamma_i = \{(x, y) | d(x, y) = i\}$ for each $i = 0, 1, \ldots, s$ (where $G \leq \mathrm{Aut}\,\Gamma$ and $d(x, y)$ denotes the distance between x and y). Then the G-symmetric graphs are precisely the $(G, 1)$-symmetric graphs, and we say that Γ is *distance transitive* if it is (G, s)-symmetric for $G = \mathrm{Aut}\,\Gamma$ and all s . Distance transitive graphs of valency 3 and 4 have been classified by Biggs and Smith [2] and Smith [19], their proofs relying heavily on the proof of Sims' conjecture for subdegrees 3 and 4 . It is a consequence of Sims' conjecture and hence of the simple group classification, see [11], that:

THEOREM 4.1 (S). *There are only finitely many distance transitive graphs of any given valency greater than* 2 .

It is hoped that the simple group classification can be used to get a much sharper result than this for distance transitive graphs. The concept of distance transitivity can be generalized as follows: a graph Γ is called *metrically n-tuple transitive* if any isometry between subsets of at most n vertices extends to an automorphism. Thus the distance transitive graphs are metrically 2-tuple transitive graphs. The metrically 6-tuple transitive graphs were characterized by Cameron [6], and using the simple group classification the metrically 5-tuple transitive graphs have been determined, see [3] and [8]. So there is quite a gap between these results and a classification of distance transitive graphs.

5. Primitive 2-transitive graphs

Let Γ be a connected G-symmetric graph of valency $v \geq 3$ such that G is primitive on vertices V . In this situation it is possible to obtain information from a result of O'Nan and Scott (see [7], [8], [1], [15]) which is a broad structure theorem for primitive permutation groups, namely:

THEOREM 5.1. *Let* G *be a primitive permutation group on a set* V . *Then one of the following four situations must arise.*

(A) G *has a unique minimal normal subgroup* N *which is a nonabelian simple group, and* $G \leq \mathrm{Aut}\,N$.

(B) G *is a group of affine transformations of a finite vector space over a prime field:* G *contains all translations and a stabilizer* G_x , $x \in V$, *is an irreducible linear group.*

(C) G *is a subgroup of the group* X *generated by* T^n , $n > 1$ *(where* T *is a nonabelian simple group), the automorphism group* $\mathrm{Aut}\,T$ *acting in the same way on each factor (where inner automorphisms are identified with the corresponding elements of the diagonal subgroup of* T^n *), and the symmetric group* S_n . *Further* V *can be taken as the set of right cosets of* $H = \mathrm{Aut}\,T \times X_n$ *in* X *with* G *(and* X *) acting* *by right multiplication.*

(D) *There is a hypercubic graph (or Hamming scheme)* $H(d, m)$, $d > 1$, $m > 1$, *with vertex set* V *which admits* G *as a group of automorphisms. (The vertices of* $H(d, m)$ *are all the* d*-tuples of elements of* $\underline{m} = \{1, 2, \ldots, m\}$ *two vertices being adjacent whenever they differ in exactly one coordinate.) Here* $G \le \mathrm{Aut}\, H(d, m) = S_m \,\mathrm{wr}\, S_d$.

In case (C), G contains T^n, and either G contains a primitive subgroup of S_n or $n = 2$ and $G \le \langle T^n, \mathrm{Aut}\, T\rangle$ (see [7], p.6, Remark 2). In case (D), if G does not also satisfy (B) then $G \le G_0 \,\mathrm{wr}\, S_d$ where G_0 is a primitive subgroup of S_m and G contains a transitive subgroup of S_d (see [7], p.5). Either G_0 satisfies (C) and G contains T^{nd}, or G_0 has a nonabelian simple normal subgroup T and $G \ge T^d$. In the latter case either T is the socle of G_0 or T is a regular normal subgroup of G_0.

In this final section of the paper we begin an investigation of primitive 2-transitive graphs Γ, of valence ≥ 3, that is connected graphs Γ with $G \le \mathrm{Aut}\,\Gamma$ such that G is primitive on the vertex set V and transitive on 2-paths. Then for a vertex x, $G^{\Gamma_1(x)}$ is 2-transitive. First we show that G cannot satisfy case (C) of Theorem 5.1.

LEMMA 5.2 (S). *If* Γ *is a connected graph of valency* $v \ge 3$, *and if* $G \le \mathrm{Aut}\,\Gamma$ *is primitive on vertices and transitive on* 2*-paths then* G *does not satisfy* (C) *of Theorem 5.1.*

Proof. Suppose that G satisfies Theorem 5.1 (C). Then V can also be taken as the set of right cosets of the diagonal subgroup $D = H \cap T^n$ of T^n in T^n, with T^n acting by right multiplication and $G \cap H$ acting by conjugation. (We shall usually choose the coset representatives ot have nth entry 1.) Taking $x = D(1, \ldots, 1)$ the stabilizer $G_x = G \cap H$ is 2-transitive on $\Gamma_1(x)$. Now the normal subgroup $M \cong T$ of G_x fixes only x (for M fixes $y = D(t_1, \ldots, t_{n-1})$ if and only if M centralizes $t_1, \ldots, t_{n-1}$, that is $t_1 = \ldots = t_{n-1} = 1$ and $y = x$). Thus $G_x^{\Gamma_1(x)}$ is a normal extension of the nonabelian simple group $M^{\Gamma_1(x)} \cong T$. Further if $y = D(t_1, \ldots, t_{n-1}) \in \Gamma_1(x)$ and $t \in M$ then $y^t = y$ if and only if t centralizes $t_1, \ldots, t_{n-1}$. Thus M_y centralizes a nontrivial element $s = t_i$ for some i, and as M_y is maximal in M (since M is primitive on $\Gamma_1(x)$) and M is simple, $s \in Z(M_y)$, that is M_y has nontrivial centre. However no 2-transitive group with a nonabelian simple normal subgroup has this property (see for example [10] for a list of such 2-transitive groups).

Next we show that if G satisfies (D) but not (B) of Theorem 5.1, then either G has a regular normal subgroup or $G_x^{\Gamma_1(x)}$ has a nontrivial abelian normal subgroup.

LEMMA 5.3 (S). *If* Γ *is a connected graph of valency* $v \ge 3$ *and if* $G \le \mathrm{Aut}\,\Gamma$

is primitive on vertices, transitive on 2-paths, and satisfies (D) *but not* (B) *of Theorem 5.1, then* $V = U^d$ *and* $N = T^d \le G \le G_0 \text{ wr } S_d$ *where* G_0 *is primitive on* U *and* T *is a nonabelian simple group. Further either*

(a) N *is regular on* V *(and* $(G \cap G_0^d)/N$ *is soluble), or*

(b) T *is the socle of* G_0 *, and if* $x = (1, \ldots, 1) \in V$, $1 \in U$;

then $\Gamma_1(x) = \Delta(1)^d$ *where* $\Delta(1)$ *is an orbit of* $(G_0)_1$ *in* U *. Moreover* $(G_0)_1^{\Delta(1)}$ *is 2-transitive with an elementary abelian regular normal subgroup* M_1 *and* $M = M_1^d$ *is an elementary abelian regular normal subgroup of* $G_x^{\Gamma_1(x)}$.

These two lemmas have an interesting corollary, namely;

COROLLARY 5.4 (S). *If* Γ *is a connected graph of valency* $v \ge 3$ *such that* $G \le \text{Aut}\,\Gamma$ *is primitive on vertices and* $G_x^{\Gamma_1(x)}$ *for a vertex* x *is 2-transitive with a nonabelian simple normal subgroup, then* G *has a unique minimal normal subgroup which either is regular on vertices, or is a nonabelian simple group. Moreover if* G_x *is not faithful on* $\Gamma_1(x)$ *then* G *has a nonabelian simple normal subgroup.*

(The last part follows from the fact that G_x is faithful on any orbit other than $\{x\}$ when G has a regular normal subgroup.) This corollary suggests that better results than Theorem 3.1 (S) and 3.2 (S) may be possible when G is primitive on vertices. Further Lemma 5.3 reduces the classification of G-symmetric graphs Γ such that G is primitive on vertices and $G_x^{\Gamma_1(x)}$ is 2-transitive with a regular normal abelian subgroup to a classification of such graphs where G has a nonabelian simple socle or a regular normal subgroup.

Proof. Suppose that G satisfies (D) but not (B) of Theorem 5.1. Then by the remark following Theorem 5.1, $G \le G_0 \text{ wr } S_d$ where G_0 is primitive on U and $V = U^d$. Either G_0 has a nonabelian simple normal subgroup T and $G \ge T^d = N$, and T is the socle of G_0 (case 1) or T is regular on U (case 2), or G_0 satisfies (C) and $G \ge T^{nd} = N$ (case 3). We shall take U as $\{1, 2, \ldots, n_0\}$ and x as $(1, \ldots, 1)$. Then G_x is 2-transitive on $\Gamma_1(x)$.

Case 1. Suppose first that N is not regular on V . Then as Γ is connected $N_x = T_1^d$ acts nontrivially on $\Gamma_1(x)$. Let $y = (b_1, \ldots, b_d) \in \Gamma_1(x)$. Without loss of generality we may assume that b_1 is not fixed by T_1 (acting on U). Let $\Delta(1)$ be the orbit of $(G_0)_1$ in U containing b_1 . Suppose that for some j, $b_j \notin \Delta(1)$. Let $a \ne b_1$ lie in the T_1-orbit containing b_1 and let $k \in G_x \cap S_d$ map j to 1 (k exists since G contains a transitive subgroup of S_d). Then $\Gamma_1(x)$ contains $z = (a, b_2, \ldots, b_d)$ and $y^k = (b_{1k^{-1}}, \ldots, b_{dk^{-1}})$ (which has first entry b_j), and G_{xy} must contain an element $g = (g_1, \ldots g_d)h$ where $g_i \in (G_0)_1$ and $h \in S_d$ which maps z to y^k . Now g fixes y if and only if for all i, $b_i = b_{ih^{-1}}^{g_i}$. Consider the first entry of $z^g = y^k$. If h fixes position 1 then we have $a^{g_1} = b_j$ which is impossible as $a^{g_1} \in \Delta(1)$. Thus h does not fix 1 and we have $b_{1h^{-1}}^{g_1} = b_j$. However $b_{1h^{-1}}^{g_1} = b_1$ is in $\Delta(1)$, again a contradiction. Thus all $b_j \in \Delta(1)$ so $\Gamma_1(x) \subseteq \Delta(1)^d$. On the other hand as G_x is 2-transitive on $\Gamma_1(x)$, $N_x = T_1^d$ is transitive on $\Gamma_1(x)$ and so T_1 is transitive on $\Delta(1)$ and $\Gamma_1(x) = \Delta(1)^d$.

Next $G_x^{\Gamma_1(x)}$ has a unique minimal normal subgroup M which contains its centralizer and is either elementary abelian or a nonabelian simple group (see [7], Prop. 5.2): M must lie in $N_x^{\Gamma_1(x)} = (T_1^{\Delta(1)})^d$. It follows that $M = M_1^d$ where M_1 is elementary abelian and regular on $\Delta(1)$. Moreover $(G_0)_1$ is 2-transitive on $\Delta(1)$: for if a, c lie in $\Delta(1)\backslash\{b_1\}$ and $z = (a, b_2, \ldots, b_d)$, $z' = (c, b_2, \ldots, b_d)$ there exists $g = (g_1, \ldots, g_d)h \in G_{xy}$ which maps z to z'. Arguing as above h fixes 1 and $g_1 \in (G_0)_1$ maps a to c and fixes b_1.

Case 2. Now suppose that $N = T^d$ is regular on V, so $V = T^d$. In this situation, as G_0 is primitive on T, G_0 is not contained in $\mathrm{Aut}\,T \times Z_2$, so G_0 contains $T \times T$. Let M be the normal subgroup of $G_0 \operatorname{wr} S_d$ isomorphic to T^{2d}. Then if we take $x = (1, \ldots, 1)$, $M \cap G_x$ is a normal subgroup of G_x. Now $M_x \simeq T^d$ acts on V by "conjugation", that is if $g = (s_1, \ldots, s_d) \in M_x$ and $y = (t_1, \ldots, t_d) \in V$ then $y^g = (s_1^{-1}t_1s_1, \ldots, s_d^{-1}t_ds_d)$. Let $y = (t_1, \ldots, t_d) \in \Gamma_1(x)$. Then $g = (s_1, \ldots, s_d) \in M$ fixes y if and only if, for all i, s_i centralizes t_i. If the subgroup induced by G on U is primitive (we can take it equal to G_0), it follows that for every $t \in T$ and every j, $1 \le j \le d$, there exists $g = (s_1, \ldots, s_d) \in M \cap G_x$ such that $s_j = t$. It follows that $M \cap G_x$ fixes y if and only if t_j is centralized by T for all $1 \le j \le d$, that is $t_1 = \ldots = t_d = 1$ and $y = x$, a contradiction. Thus $M \cap G_x$ does not fix y and so $(M \cap G_x)^{\Gamma_1(x)}$ is a nontrivial normal subgroup of $G_x^{\Gamma_1(x)}$. It follows that $(M \cap G_x)^{\Gamma_1(x)} \simeq T$ and $T \le G_x^{\Gamma_1(x)} \le \mathrm{Aut}\,T$. Further our argument above showed that $M \cap G_x$ acts as T on each component so that $(M \cap G_x)^{\Gamma_1(x)}$ acts as a diagonal subgroup of T^d. If $y = (t_1, \ldots, t_d) \in \Gamma_1(x)$ then some entry t_j is not the identity, and $(M \cap G_{xy})^{\Gamma_1(x)} \le C_T(t_j)$. Then as $(M \cap G_{xy})^{\Gamma_1(x)}$ is a maximal subgroup of T (since $(M \cap G_x)$ is primitive on $\Gamma_1(x)$), and as T is simple, it follows that t_j is in the centre of $(M \cap G_{xy})^{\Gamma_1(x)}$. (As in Lemma 5.2), no 2-transitive group with a nonabelian simple normal subgroup has this property. Thus the subgroup induced by G on U is not primitive so that $M \cap G_x$ is trivial and $G_x \cap G_0^d \le H^d$ where H is a subgroup (possibly trivial) of $\mathrm{Aut}\,T$ which avoids T, and hence is soluble.

Case 3. Suppose that G_0 satisfies (C) of Theorem 5.1, so $G \ge T^{nd} = N$. Here

$x = (1, \ldots, 1)$ where $1 = D(1, \ldots, 1)$, D the diagonal subgroup of T^n. Let $y = (b_1, \ldots, b_d) \in \Gamma_1(x)$ where each b_i is a coset of D in T^n. Since G_x contains a transitive subgroup of S_d there is for each $i = 1, \ldots, d$ such a point y in $\Gamma_1(x)$ with ith entry $b_i \neq 1$, and hence $N_x \simeq T^d$ acts faithfully on $\Gamma_1(x)$. This is impossible as a 2-transitive group has a simple or abelian normal subgroup which contains its centralizer. This completes the proof of Lemma 5.3.

References

[1] M. Aschbacher and L.L. Scott, "Maximal subgroups of finite groups", *J. Algebra*, to appear.

[2] N.L. Biggs and D.H. Smith, "On trivalent graphs", *Bull. London Math. Soc.* 3 (1971), pp. 155-158, MR44:3902.

[3] J.M.J. Buczak, "Finite group theory", D.Phil. Thesis, Oxford University (1980).

[4] P.J. Cameron, "Permutation groups with multiply transitive suborbits", *Proc. London Math. Soc.* (3) 25 (1972), pp. 427-440, MR47:5082.

[5] P.J. Cameron, "Suborbits in transitive permutation groups", *Combinatorics* (eds. M. Hall, Jr and J.H. van Lint, Math. Centre Tracts 57, Math. Centrum, Amsterdam, 1973), pp. 98-129, MR51:5718.

[6] P.J. Cameron, "6-Transitive graphs", *J. Combinatorial Theory (B)* 28 (1980), pp. 168-179, MR81g:05076.

[7] P.J. Cameron, "Finite permutation groups and finite simple groups", *Bull. London Math. Soc.* 13 (1981), pp. 1-22, MR80m:20008.

[8] P.J. Cameron, "Finite simple groups and finite geometries", *Proceedings of Pullman Conference on Finite Geometry* (1981), to appear.

[9] P.J. Cameron and C.E. Praeger, "Graphs and permutation groups with projective subconstituents", *J. London Math. Soc.* (2) 25 (1982), pp. 62-74.

[10] P.J. Cameron and C.E. Praeger, "On 2-arc transitive graphs of girth 4", *J. Combinatorial Theory (B)* 35 (1983), pp. 1-11.

[11] P.J. Cameron, C.E. Praeger, J. Saxl, and G.M. Seitz, "On the Sims conjecture and distance transitive graphs", *Bull. London Math. Soc.* 15 (1983), pp. 499-506.

[12] A. Gardiner, "Arc transitivity in graphs", *Quart. J. Math. Oxford* (2) 24 (1973), pp. 399-407, MR48:1973.

[13] A. Gardiner, "Symmetry conditions in graphs", *Surveys in Combinatorics* (ed. B. Bollabas, London Math. Soc. Lecture Notes in Math. 38, Cambridge University Press, Cambridge, 1979), pp. 22- 43, MR81e:05081.

[14] D.G. Higman, "Intersection matrices for finite permutation groups", *J. Algebra* 6 (1967), pp. 22-42, MR35:244.

[15] L.G. Kovács, "Maximal subgroups in composite finite groups", (unpublished).

[16] C.E. Praeger, "Primitive permutation groups and a characterization of the odd graphs", *J. Combinatorial Theory (B)* 31 (1981), pp. 117-142, MR83c:20003.

[17] C.E. Praeger, "When are symmetric graphs characterised by their local

properties?", *Combinatorial Mathematics* IX (eds. E.J. Billington, S. Oates-Williams, and A.P. Street, Lecture Notes in Math. 952, Springer, Berlin-Heidelberg-New York, 1982), pp. 123-141.

[18] C.E. Praeger, "Finite simple groups and finite primitive permutation groups", *Bull. Austral. Math. Soc.* 28 (1983), pp. 355-366.

[19] D.H. Smith, "Distance-transitive graphs of valency four", *J. London Math. Soc.* (2) 8 (1974), pp. 377-384, MR52:2026.

[20] J.G. Thompson, "Bounds for the order of maximal subgroups", *J. Algebra* 14 (1970), pp. 135-138, MR40:5720.

[21] W.T. Tutte, "A family of cubical graphs", *Proc. Camb. Phil. Soc.* 43 (1947), pp. 459-474, MR9,p7.

[22] W.T. Tutte, "On the symmetry of cubic graphs", *Canad. J. Math.* 11 (1959), pp. 621-624, MR22:679.

[23] R. Weiss, "s-transitive graphs", *Algebraic Methods in Graph Theory* (Colloquia Math. Soc. Janos Bolyai 25, North-Holland, Amsterdam, 1981), pp. 827-847, MR83b:05071.

[24] H. Wielandt, *Finite Permutation Groups* (Academic Press, New York-London, 1964), MR32:1252.

Department of Mathematics,
University of Western Australia,
Nedlands, W.A. 6009
Australia

PROCEEDINGS OF 'GROUPS – KOREA 1983'
KYOUNGJU, August 1983 20F10, 20F16

DECISION PROBLEMS FOR INFINITE SOLUBLE GROUPS

Derek J.S. Robinson

1. Introduction

Let G be a group with a presentation $\langle X|R\rangle$ where X is a countable set. We shall be concerned with three decision problems for the presentation.

a) *The word problem* (w.p.). Is there an algorithm to decide if an element of G (given as a word in X) is the identity?

b) *The generalized word problem* (g.w.p.). Is there an algorithm to decide membership of elements of G in a given r.e. subgroup of G ?

c) *The conjugacy problem* (c.p.). Is there an algorithm to decide if two elements of G are conjugate?

Clearly the truth of b) or c) implies that of a). Also a necessary condition for a) to hold is that the presentation be recursive (i.e., the set of all relations is r.e.).

We shall say that G has soluble w.p., g.w.p. or c.p. if the problem is soluble for some presentation of G .

Some known results

It is known that all three problems have negative solutions for finitely presented soluble groups. In fact

i) *There is a finitely presented soluble group in the variety* $N_4A \cap A_2A_2A$ *which has insoluble word problem* (Harlampovič [5]).*

(Here N_c is the class of nilpotent groups of class $\leqslant c$ and A and A_2 are respectively the classes of abelian groups and elementary abelian 2-groups).

ii) *All three problems have positive solutions for polycyclic groups.*

This follows from results of Mal'cev [8] (polycyclic groups have separable subgroups) and of Formanek [3] and Remeslennikov [9] (polycyclic groups are conjugacy separable).

iii) *A finitely generated soluble group with finite rank has soluble w.p. if and*

* Further examples have been given by Baumslag, Gildenhuys

only if it is recursively presented (Cannonito and Robinson [2]).

This means that every finitely presented soluble group of finite rank has soluble w.p.

iv) *A group in the class* NPF *which satisfies* max-n *has soluble* w.p. (Baumslag, Cannonito, and Miller [1]).

Here P and F denote the classes of polycyclic groups and finite groups respectively. For example, it follows from (iv) and a well-known theorem of P. Hall that all finitely generated APF-groups have soluble w.p.

On the other hand, we remark that not all soluble groups with max-n have soluble w.p.

v) *There is a 3-step soluble group with* max-n *which does not have a recursive presentation and so has insoluble* w.p.

We shall describe the construction briefly; it is a variant of an idea of Hall [4].

Construction

Let p be any prime and write $\mathbb{Q}_p$ for the ring of rational numbers with denominator a power of p. Let V be a free $\mathbb{Q}_p$-module with basis $\{v_n | n \in \mathbb{Z}\}$. Choose an injection $\lambda : \mathbb{Z} \to \mathbb{Z}$. We define two automorphisms α, β of V by the rules

$$v_n\alpha = p^{\lambda(n)}v_n \quad \text{and} \quad v_n\beta = v_{n+1}.$$

It is easy to see that $Q = \langle\alpha, \beta\rangle$ is the wreath product of two infinite cyclic groups, provided that λ does not satisfy a homogeneous linear difference equation with integral coefficients.

We claim that V *is a noetherian Q-module*; to establish this we take a non-zero Q-submodule U of V and choose u a non-zero element of U which has minimal length as a linear combination of basis elements, say

$$u = l_1v_{i_1} + \ldots + l_kv_{i_k}$$

where $0 \neq l_i \in \mathbb{Q}_p$ and $i_1 < \ldots < i_k$. Now U will also contain the element

$$u(\alpha - p^{\lambda(i_1)}) = \sum_{r=2}^{k} l_r(p^{\lambda(i_r)} - p^{\lambda(i_1)})v_{i_r}.$$

Since λ is injective, the minimality of k shows that $k = 1$ and $l_1v_{i_1} \in U$. From this it is easy to see that $U \geqslant mV$ for some positive integer m which is prime to p.

But V/mV is isomorphic as a $\langle\beta\rangle$-module with the group ring $\mathbb{Z}_m\langle\beta\rangle$, which is a noetherian ring. Hence V/mV is a noetherian Q-module and our claim is established.

Next we form the semidirect product

$$G(p, \lambda) = Q \ltimes A .$$

Since Q satisfies max-n, it follows from the previous paragraph, that $G(p, \lambda)$ too satisfies max-n. Of course it is clear that $G(p, \lambda)$ is 3-step soluble. To complete the proof of (v) we shall prove

vi) *If* λ *is computable, the group* $G \equiv G(p, \lambda)$ *has soluble* w.p. *Conversely, if* G *has a recursive presentation,* λ *is computable.*

Proof. As a first step note that G has the following finitely generated presentation:

$$\text{generators: } \alpha, \beta, v_0$$

$$\text{relations: } [\alpha, \alpha^{\beta^i}] = 1 \quad (i \in \mathbb{Z}) ,$$

$$v_{ij}^{\alpha^{-1}} = v_{ij+\lambda(i)} , \quad v_{ij}^{\beta} = v_{i+1j} ,$$

$$v_{ij+1}^{p} = v_{ij} , \quad [v_{ij}, v_{kl}] = 1 ,$$

where v_{ij} is defined by the rule

$$v_{ij} = ((v_0^{\beta^i})^{\alpha^{-j}})^{p^{j(\lambda(i)-1)}} \qquad (j \geq 0) .$$

If λ is computable, there is an effective way of writing an arbitrary word g in α, β, v_0 in the form

$$\beta^n \alpha_{l_1}^{m_1} \ldots \alpha_{l_r}^{n_r} v_{i_1 j_1}^{k_1} \ldots v_{i_s j_s}^{k_s}$$

where $\alpha_t = \alpha^{\beta^t}$, $l_1 < \ldots < l_r$, $i_1 < \ldots < i_s$, $m_t \neq 0$ and $p \nmid k_t$. Then $g = 1$ if and only if $n = r = s = 0$. Hence the w.p. is soluble for this presentation.

Conversely, suppose that G has a recursive presentation. Then the given presentation is also recursive. Enumerate all relations in the given presentation and at the same time enumerate for $t = 0, 1, 2, 3, \ldots$ all words of the form

$$(v_0^{\beta^i})^{\alpha}(v_0^{\beta^i})^{-p^t} .$$

Exactly one such word will be a relation for each i. We determine this t and put $\lambda(i) = t$. Hence λ is computable.

In view of (iv) and (vi) it is natural to raise the following:

Question. *If a soluble group with* max-n *has a recursive presentation, does the group have soluble* w.p.?

In particular one can ask if a finitely presented soluble group with max-n has soluble w.p. The corresponding question for min-n has a positive answer even if the group is insoluble. This follows from a result of Huber-Dyson [6].

2. Decision problems for infinitely generated soluble groups of finite rank

In the second section we shall describe some new results applicable to soluble groups of finite rank that are not necessarily finitely generated. First of all we shall recall the principal classes of soluble groups of "finite rank". These classes are listed in the accompanying diagram; all groups are soluble.

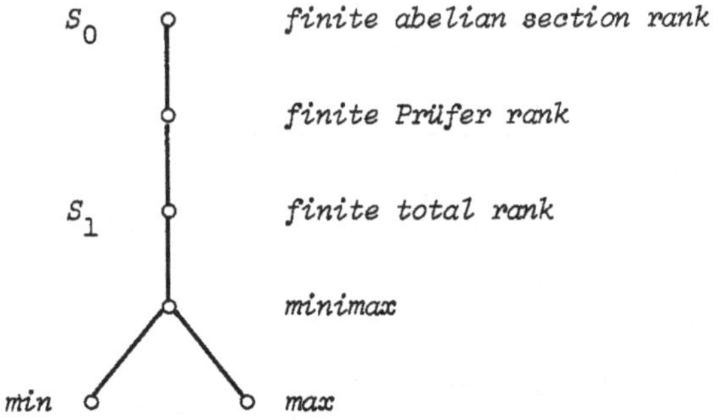

It is known that the top four classes coincide for finitely generated groups (Robinson [10]).

These are three main results on the decision problems (a), (b), (c).

THEOREM A. *Let G be a soluble group with finite total rank. If G has a recursive presentation, then the word problem is soluble for that presentation.*

THEOREM B. *Let G be a soluble minimax group with a recursive presentation. If H is a subgroup which is r.e. in terms of the presentation, there is an algorithm to decide membership in H.*

THEOREM C. *Let G be a soluble minimax group with a recursive presentation. If g is a fixed element of G, there is an algorithm to decide if an arbitrary element is conjugate to g.*

In the sense of these results we may conclude that a soluble minimax group which is finitely presented in some variety A^l has soluble w.p., g.w.p. and c.p. The same conclusion applies to finitely generated soluble groups that are residually finite since such groups are recursively presented (see [2]).

Three examples

We mention three examples which limit the validity of results such as Theorems A, B, C.

a) *Theorem A is false for soluble groups of finite Prüfer rank.*

Let π be a r.e., non-recursive set of primes. Define G to be the abelian group with generators x_p where p is prime, subject to relations $x_p^p = 1$, and also $x_p = 1$ if $p \in \pi$. Clearly G is a direct product of groups of order p where $p \in \pi'$, so G has Prüfer rank 1; also G has a recursive presentation. However the w.p. is insoluble for every presentation; for otherwise we could determine the orders of the generators and so enumerate the primes in π'.

b) *Theorem B is false for the class* S_1.

Let π be as in a) and let S be the additive group of all square-free rational numbers. Then $T = \left\langle \frac{1}{p} \mid p \in \pi \right\rangle$ is a r.e. subgroup of S and $S/T \simeq G$. We cannot decide membership in T since the w.p. is insoluble in G.

c) *Theorem C is false for the class* S_1.

Let S and T be as in b). Define $A = Z \oplus S$. If $t \in T$, the matrix

$$\xi_t = \begin{pmatrix} 1 & t \\ 0 & 1 \end{pmatrix}$$

determines an automorphism of A. Let $X = \{\xi_t \mid t \in T\}$; this subgroup of Aut A is isomorphic with T. Let H be the semidirect product

$$X \ltimes A .$$

Then H is an S_1-group with a recursive presentation. Let $h = (1, 0)$ and $h_s = (0, s)$, where $s \in S$. Then h_s is conjugate to h in H if and only if $s \in T$. Therefore we cannot decide if h_s is conjugate to h.

Thus there is a clear difference between soluble minimax groups and groups in the class S_1.

Remark. We cannot approach Theorems B and C by appealing to separability of subgroups or conjugacy separability, as was done for polycyclic groups (see §1 (ii)). The reason is that a finitely presented soluble minimax group need not have either of these properties. Indeed Wehrfritz [11] has proved that the group

$$G = \langle a, b, x, y \mid [a, b] = 1 , \quad a^x = a, \ b^x = a^2 b, \ a^y = a^2 ,$$

$$b^y = b^2 , \quad [x, y] = 1 \rangle$$

is not conjugacy separable, while it follows from a theorem of Jeanes and Wilson [7] that G does not have separable subgroups.

3. Sketch of proofs

We shall give brief accounts of the proofs of Theorems A and B. The proof of Theorem C is much more complicated and we cannot usefully say anything about it here.

Proof of Theorem A. The basic idea derives from [6]. Let $R \rightarrowtail F \xrightarrow{\pi} G$ be a recursive presentation of the group G where F is a free group. Since $G \in S_1$, there is a direct product $D = G_1 \times \dots \times G_n$ where $1 \neq G_i \triangleleft G$ and G_i is abelian and either torsion-free or an elementary abelian p-group; moreover n is to be chosen as large as possible. If $1 \neq N \triangleleft G$, then in fact $N \cap D \neq 1$. This permits the detection of non-relations.

In each G_i choose a maximal independent subset and take pre-images under π. Let the resulting finite subset of F be $\{u_1, \dots, u_r\}$.

Let $w \in F$; then either $w \in R$ or else $\langle w^F \rangle^\pi \cap D \neq 1$. We adopt the following procedures.

i) Enumerate R.

ii) Enumerate the set $S \cap R$; here S is the r.e. set of all words

$$vu_{i_1}^{l_1} \dots u_{i_t}^{l_t}$$

where $t > 0$, $i_1 < i_2 < \dots < i_t$, $v \in \langle w^F \rangle$ and $l_i \neq 0$ if u_i^π has infinite order while $l_i \not\equiv 0 \bmod p_i$ if u_i^π has prime order p_i.

Either we shall find w in R or else we shall find an element of $S \cap R$, in which case w 1. Hence the w.p. is soluble for the given presentation.

Proof of Theorem B. Let d denote the derived length of G. If $d \leqslant 1$, then G is abelian and $H \triangleleft G$. The group G/H is minimax and it has a recursive presentation since H is r.e. By Theorem A the w.p. is soluble for G/H; thus we can decide membership in H.

Let $d > 1$ and proceed by induction on d. Put $D = G^{(d-1)}$. Then G/D has a recursive presentation, so we can decide membership in the subgroup HD/D. Let x be an element of G. We decide if xD belongs to HD/D. If not, then $x \notin H$, so assume that $x \in HD$. Now enumerate the elements $h^{-1}x$, $h \in H$, and check to see if $h^{-1}x$ belongs to D. This is possible by Theorem A because G/D is a soluble minimax group with a recursive presentation. In this way we can find an h in H such that $d \equiv h^{-1}x$ belongs to D. Finally $D/H \cap D$ has soluble w.p. because $H \cap D$ is r.e., so we can decide if d belongs to $H \cap D$. Clearly $x \in H$ if and only if $d \in H \cap D$.

References

[1] G. Baumslag, F.B. Cannonito, and C.F. Miller III, "Some recognizable properties of solvable groups", *Math. Z.* 178 (1981), pp. 289-295, MR82k:20061.

[2] F.B. Cannonito and D.J.S. Robinson, "The word problem for finitely generated soluble groups of finite rank", *Bull. London Math. Soc.* 16 (1984), pp. 43-46.

[3] E. Formanek, "Conjugacy separability in polycyclic groups", *J. Algebra* 42 (1976), pp. 1-10, MR54:7626.

[4] P. Hall, "On the finiteness of certain soluble groups", *Proc. London Math. Soc.* (3) 9 (1959), pp. 595-622, MR22:1618.

[5] O.G. Harlampovič, "A finitely presented soluble group with insoluble word problem", *Izv. Akad. Nauk SSSR* 45 (1981), pp. 852-873, 928 (Russian), MR82m:20036.

[6] V. Huber-Dyson, "Finiteness conditions and the word problem", *Groups — St. Andrews 1981* (London Math. Soc. Lecture Notes 71, 1982), pp. 244-251.

[7] S.C. Jeanes and J.S. Wilson, "On finitely generated groups with many profinite-closed subgroups", *Arch. Math. (Basel)* 31 (1978), pp. 120-122, MR80b:20038.

[8] A.I. Mal'cev, "On homomorphisms into finite groups", *Učen. zap. Ivanov. ped. inst.* 18 (1958), pp. 49-60 (Russian).

[9] V.N. Remeslennikov, "Conjugacy in polycyclic groups", *Algebra i Logika* 8 (1969), pp. 712-725 (Russian), MR43:6313.

[10] D.J.S. Robinson, "On the cohomology of soluble groups of finite rank", *J. Pure Appl. Algebra* 6 (1975), pp. 155-164, MR52:3363.

[11] B.A.F. Wehrfritz, "Two examples of soluble groups that are not conjugacy separable", *J. London Math. Soc.* (2) 7 (1973), pp. 312-316, MR49:2942.

Department of Mathematics,
University of Illinois,
Urbana-Champaign, Illinois 61801
U.S.A.

PROCEEDINGS OF 'GROUPS — KOREA 1983' 16A26, 20C05
KYOUNGJU, August 1983

AUTOMORPHISMS AND ISOMORPHISMS OF INTEGRAL GROUP RINGS OF FINITE GROUPS

K.W. Roggenkamp

This is a *preliminary report* of joint work with L.L. Scott.

The problem we are considering is the notorious "Isomorphism problem": If for two finite groups G and H the integral group rings $\mathbb{Z}G$ and $\mathbb{Z}H$ are isomorphic, are then necessarily G and H isomorphic. The problem is of interest not so much that a solution will severely influence integral representation theory, but since it has for over 40 years resisted many attacks.

For the sake of simplicity we formulate our results on p-groups only for $\hat{\mathbb{Z}}_p$, the p-adic integers, though they hold for finite unramified extensions.

1. Introduction and notation

If R is a commutative ring with identity and G a finite group, we denote by $U(RG)$ *the units in the group ring* RG. The *augmentation map* $\varepsilon_G : RG \to R$ is induced by sending $g \mapsto 1$ and has as kernel the *augmentation ideal* $I(RG)$. We denote by $V(RG)$ the *normalized units*, the units of augmentation 1, i.e. $V(RG) = U(RG) \cap (1 + I(RG))$. We recall that an isomorphism $\phi : RG \to RH$ can easily be modified to yield an *augmented isomorphism* $\phi_\varepsilon : RG \to RH$, i.e. $\varepsilon_G = \phi_\varepsilon \varepsilon_H$.

We next assume that R is *G-adapted*, i.e. R is an integral domain, with

(i) char $R = 0$,

(ii) no prime divisor of $|G|$, the order of G, is a unit in R.

With this notation we have the following equivalent formulation of the isomorphism problem: G and H are finite groups and R is G-adapted:

IP: Does $RG = RH$ imply $G = H$?

AIP: Does $RG = RH$ as augmented rings: i.e. $I(RG) = I(RH)$ imply $G = H$?

UP: Let U be a finite subgroup of $V(RG)$ with $|U| = |G|$. Is then $U \simeq G$?

To see the implication UP $\Rightarrow$ AIP one has to invoke the result of G. Higman [5] which says that in UP, the elements in U are "linearly independent" over R in RG.

In our approach to the problem we have concentrated on UP. In connection with UP there are conjectures of Zassenhaus [17]:

Z1: Let U be a finite subgroup of $V(RG)$ with $|U| = |G|$. Then there exists a unit $a \in KG$, K the field of fractions of R, such that

$$aUa^{-1} = G .$$

This obviously implies UP, but it is much stronger: For example it says that the action of a normalized automorphism α of RG on the centre $Z(RG)$ is induced from a group automorphism of G.

An even stronger conjecture — also more or less due to Zassenhaus [17] — is:

Z2: Let U be a cyclic subgroup of $V(RG)$, then there exists a unit $a \in KG$ such that

$$aUa^{-1} \subset G .$$

The conjecture Z2 has recently been verified by Ritter and Sehgal [11] in the following cases

(i) $G = \langle a \rangle] \langle x \rangle$, $a^{p^m} = x^t = 1$, $(p, t) = 1$, p prime

(ii) $G = \langle a \rangle] \langle x \rangle$, $a^n = x^p = 1$, n odd, p prime

(iii) G nilpotent of class 2.

But in all these cases a positive answer to the isomorphism problem is known since these groups are metabelian and we have

THEOREM 1 (Jackson [6], Whitcomb [16]). *Let G be metabelian and R a commutative ring which is G-adapted, then* AIP *has a positive solution.*

The proof, which we shall analyze further down, even shows that the "metabelian" extension classes of G and H are isomorphic.

When Leonard Scott and I worked on this problem we became sooner or later — in this case later — convinced that it was hard to find a counter-example for nilpotent — even solvable — groups. So we were looking for attacks in the positive direction, and influenced by several examples and the proof of Theorem 1 we were looking for *some kind of induction.* None of the equivalent formulations of the isomorphism problem or the Zassenhaus conjectures is suitable for induction.

However, a much stronger property than Z1 can be used for induction, provided the ring R has additional properties. We shall call this the *trivial automorphism property of RG*.

TA*: Let U be a finite subgroup of $V(RG)$ with $|U| = |G|$. Then there

* May 1984: In the meantime we have proved TA for all p-groups and IP for nilpotent groups.

exists a unit $v \in RG$ such that

$$v\,U v^{-1} = G\ .$$

REMARKS 1. 1) Obviously TA for R implies ZI for R .

2) TA is definitely not true for $R = \mathbb{Z}$; e.g. the dihedral group of order 8 does not satisfy TA (cf. below).

3) The name "trivial automorphism-property" is justified by the following observation: Denote by Aut $N(RG)$ the *normalized automorphisms* of RG and by Inn(RG) the *inner automorphisms* of RG — these are obviously normalized. Aut(G) — the automorphisms of G embed into Aut $N(RG)$ by linear extension. Then TA is equivalent to Aut $N(RG)$ = Aut G . Inn(RG) , as is easily seen.

THEOREM 2 (Fröhlich [2], Endo-Miyata-Sekiguchi [1], Sekiguchi [15], Roggenkamp-Scott). *Let* $R = \hat{\mathbb{Z}}_p$ *the p-adic completion of the rational integers.*

(i) $H_{4.2}$ *, the quaternions of order* 8 *have* TA *for* $\hat{\mathbb{Z}}_2$ [2].

(ii) $D_{2^n.2}$ *the dihedral groups of order* $2^n.2$ *have* TA *for* $\hat{\mathbb{Z}}_2$ [1].

(iii) $G = \{a, b : a^{p^n} = b^{p^m} = 1\ ,\ b\,a\,b^{-1} = a^{1+p^r}\}$, p *odd and* $n \leqslant r+m$, $r \leqslant n \leqslant 2r$ *have* TA *for* $\hat{\mathbb{Z}}_p$ [14].

(iv) The quaternion groups of order 2^n *have the property* TA *for* $\hat{\mathbb{Z}}_2$.†

The proof in all these cases is done by explicitly writing down the p-adic group ring and then checking that there are no exceptional automorphisms; a difficult task which cannot be expected to give a general argument.

We can prove

THEOREM 3. *Let* G *be a nilpotent class* 2 *p-group, then* G *has the property* TA *for* $R = \hat{\mathbb{Z}}_p$. *This is a consequence of a more general result: We say that a p-group* G *satisfies the*

Hypothesis (*). There is C , a central subgroup of order p in G , such that there are no elements $y_1, \dots, y_n$ in G with $C_G(y_i) = C_G(Cy_i)$ and there are

$$\gamma_i \in H^1(C_G(y_i)/Z,\ F_p)$$

with

$$\gamma_i{\uparrow}^{G/Z}(y_{i+1}) \neq 0 \qquad \text{with} \qquad y_{n+1} = y_1$$

where Z is the centre of G , and $\gamma_i{\uparrow}^{G/Z}$ is the transfer from $C_G(y_i)/Z$ to G/Z

There do not seem to be too many groups which do not satisfy (*) . The first

---

† The restriction here seems to be only such that the calculations which Sekiguchi does can be carried through. In fact we have for example shown that $\langle a^{p^3}, b^{p^2}\rangle$ is o.k.

one we could find, is constructed as follows: Let $H = \langle a, b : a^{16} = b^4 = 1, a^b = a^5 \rangle$ and put $H_1 = H \times H$ and $G = H_1 \text{ wr } C_2$, $(|G| = 2^{25})$.

THEOREM 4. *Assume that* G *satisfies the hypothesis* (*) *and* G/C *has* TA *for* $\hat{\mathbb{Z}}_p$, *then* G *has* TA *for* $\hat{\mathbb{Z}}_p$.

The techniques involved in the proof of Theorem 4 seems to generalize, since they do not need the whole structure of the p-adic group ring. We shall elaborate on that in §5.

THEOREM 5.[†] *Let* $G = P_1 \times \ldots \times P_n$ *be a direct product of* p_i*-groups,* $p_i \neq p_j$ *for* $i \neq j$. *Assume that each* P_i *and all of its quotients have* TA *for unramified extensions of* $\hat{\mathbb{Z}}_{p_i}$. *Then* G *has* TA *for* $Z_\pi = \bigcap_{p \in \pi} \mathbb{Z}_p$, $\mathbb{Z}_p$ *denoting the localization at* p, π *a finite set of primes,* $p_i \in \pi$.

2. What "TA" does for the isomorphism problem

In this section we shall extend Whitcomb's proof of the isomorphism problem in the metabelian case — as presented in [9] — to "abelian extensions by TA-groups". For a finite group G we denote by $\pi(G)$ the set of rational prime divisors of $|G|$. Moreover, for a finite set of rational primes π, let $\mathbb{Z}_\pi = \bigcap_{p \in \pi} \mathbb{Z}_p$ be the π-adic localization of $\mathbb{Z}$.

THEOREM 6. *Let* G *be a finite group which satisfies* TA *for the ring* $\mathbb{Z}_\pi$, $\pi = \pi(G)$. *If*

$$0 \to A \to E \to G \to 1$$

is a group extension with A *abelian, then* E *is determined by its integral group ring; i.e. the isomorphism problem for* E *has a positive solution.*

REMARKS 2. 1) Because of the Noether-Deuring theorem for extension categories [12] the conclusion of Theorem 6 remains valid if R is a finite extension of $\mathbb{Z}_\pi$ and G satisfies TA for R.

2) In view of Theorem 3 this gives a positive answer to the isomorphism problem for nilpotent groups of class at most 5.

Before we come to sketch the proof of Theorem 6, we have to recall some *equivalence of categories* — namely of extension categories [4], [9]. However, we shall formulate the result only for extension groups.

PROPOSITION 1. *Let* R *be a commutative* G*-adapted ring and* A *a finite abelian group such that* $R \otimes_{\mathbb{Z}} A \overset{nat.}{\cong} A$. *Then the following extension groups are naturally isomorphic*

(i) $H^2(G, A)$,

---

† May 1984: This statement is not correct; but that does not affect the other results.

(ii) $\mathrm{Ext}^1_{RG}(I(RG), A)$,

(iii) $\mathrm{Ext}^1_{R\text{-alg}}(RG, A)$, the extensions of R-algebras, where A is viewed as ring with trivial multiplication.

Indication of the proof — on the level of extensions — not just of equivalence classes of extensions: Given a group extension

$$1(E): \qquad 0 \to A \to E \to G \to 1 ,$$

we have the group ring extensions

$$2(E): \qquad 0 \to I(RA)\otimes_{RA}RE \to RE \to RG \to 0 ,$$

which gives by restriction to $I(RE)$ the exact sequence

$$3(E): \qquad 0 \to I(RA)\otimes_{RA}RE \to I(RE) \to I(RG) \to 0 .$$

Because of our hypotheses, we have a natural — in particular G-equivariant — homomorphism — H. Hopf:

$$\kappa: \qquad I(RA)\otimes_{RA}RE \to A \quad \text{induced by} \quad a-1 \mapsto a ,$$

which has kernel $I(RA)I(RG)$.

Hence from $2(E)$ we get the exact sequence of R-algebras — we call this the *small isomorphism sequence*.

$$4(E): \qquad 0 \to A \to RG/I(RA)I(RG) \to RG \to 0 ,$$

we also get from $3(E)$ the exact sequence of RG-modules.

$$5(E): \qquad 0 \to A \to I(RE)/I(RA)I(RG) \to I(RG) \to 0 .$$

Going from $4(E)$ to $1(E)$ is done by forming the pullback along the natural embedding $G \to RG$. □

REMARKS 3. 1) The proof of the isomorphism problem uses not the full strength of the hypothesis that if for two metabelian groups $0 \to A \to E_i \to G \to 0$, $i = 1, 2$, the group rings $\mathbb{Z}E_1$ and $\mathbb{Z}E_2$ are isomorphic, then both extensions are equivalent. It only uses that the corresponding small isomorphism sequences are equivalent to show that the group extensions are equivalent. So we can add a new problem to our list, the *small isomorphism problem;*

SIP: Let E_1 and E_2 be given as group extensions $1(E_1)$, $1(E_2)$ with A abelian. Assume the small isomorphism sequences $4(E_1)$ and $4(E_2)$ are equivalent. Are then $1(E_1)$ and $1(E_2)$ equivalent?

2) As noted for metabelian groups IP for $\mathbb{Z}$ is proved by proving SIP for $\mathbb{Z}$

and I do not know of any counterexample over $\mathbb{Z}$ to SIP.

3) It is still an open question where IP is true for p-groups and R a field of characteristic p. There are — as yet unpublished — examples of W. Kimmerle and L. Scott which show that SIP is not true for p-groups over a field of characteristic p. However, there are no such examples over $\hat{\mathbb{Z}}_p$.

Before we come to the proof of Theorem 6, we need another easy observation.

PROPOSITION 2. *Let G be a finite group which has* TA *for* $\hat{\mathbb{Z}}_\pi$, $\hat{\mathbb{Z}}_\pi$, $\pi = \pi(G)$. *Then G has* Z1 *for* $\mathbb{Z}$ *(i.e. the stronger version π-adically implies the weaker one globally).*

Proof. Let U be a finite subgroup of $V(\mathbb{Z}G)$ with $|U| = |G|$. Since the elements in U are $\mathbb{Z}$-linearly independent, we conclude of [5] $\mathbb{Z}U = \mathbb{Z}G$ as augmented rings. Moreover, by TA and since U can also be viewed as subgroup of $V(\hat{\mathbb{Z}}_\pi G)$ there exists a unit v in $\hat{\mathbb{Z}}_\pi G$ with

$$v\,Uv^{-1} = G\ .$$

Abstractly, this is an isomorphism $\phi : U \to G$, which induces a normalized automorphism — also denoted by ϕ —

$$\phi : \mathbb{Z}U \to \mathbb{Z}G\ .$$

Since π-adically ϕ is conjugation by v, ϕ leaves the centre of $\mathbb{Z}G = \mathbb{Z}U$ elementwise fixed, and so it is conjugation by a unit in $\mathbb{Q}G$; whence Z1. □

Proof of Theorem 6. If $A = \oplus A_p$ is the decomposition of A into its p-primary components, then this is at the same time a decomposition of A as G-module. Because of the functoriality in Proposition 1 it thus suffices to treat only the cases where A is a p-group.

Case 1: p is in π. Then we can use Proposition 1 for $R = \hat{\mathbb{Z}}_\pi$. By the usual techniques, the small group ring extensions are the same

$$(*)\qquad \begin{array}{ccccccccc} 0 \to & A_1 & \to & \hat{\mathbb{Z}}_\pi E_1/I(\hat{\mathbb{Z}}_\pi E_1)I(\hat{\mathbb{Z}}_\pi A_1) & \to & \hat{\mathbb{Z}}_\pi G_1 & \to 1 \\ & \wr\Vert & & \wr\Vert & & \wr\Vert & \\ 0 \to & A_2 & \to & \hat{\mathbb{Z}}_\pi E_2/I(\hat{\mathbb{Z}}_\pi E_2)I(\hat{\mathbb{Z}}_\pi A_2) & \to & \hat{\mathbb{Z}}_\pi G_2 & \to 1\ . \end{array}$$

According to the hypothesis TA there exists a unit $v'' \in V(\hat{\mathbb{Z}}_\pi G_1)$ with $v''G_2v''^{-1} = G_1$. Since $\operatorname{rad} \hat{\mathbb{Z}}_\pi E_1/I(\hat{\mathbb{Z}}_\pi E_1)I(\hat{\mathbb{Z}}_\pi A_1) \supset A_1$, we can lift v'' to a unit $v \in \hat{\mathbb{Z}}_\pi E_1/I(\hat{\mathbb{Z}}_\pi E_1)I(\hat{\mathbb{Z}}_\pi A_1) =: \widetilde{\hat{\mathbb{Z}}_\pi E_1}$. If ϕ'' is conjugation with v'' and ϕ is conjugation with v, we have the commutative diagram with exact rows

$$\begin{array}{ccccccccc} 0 & \to & A_1 & \to & \widetilde{\hat{\mathbb{Z}}_\pi E_1} & \to & \hat{\mathbb{Z}}_\pi G_1 & \to & 1 \\ & & \uparrow\phi' & & \uparrow\phi & & \uparrow\phi'' & & \\ 0 & \to & A_2 & \to & \widetilde{\hat{\mathbb{Z}}_\pi E_2} & \to & \hat{\mathbb{Z}}_\pi G_2 & \to & 1 . \end{array}$$

Since $\phi'' : G_2 \to G_1$ we may use Proposition 1 to conclude that we get a commutative diagram of group extensions

$$\begin{array}{ccccccccc} 0 & \to & A_1 & \to & E_1 & \to & G_1 & \to & 1 \\ & & \uparrow & & \uparrow & & \uparrow & & \\ 0 & \to & A_2 & \to & E_2 & \to & G_2 & \to & 1 , \end{array}$$

which is what we wanted to prove.

Case 2: p is not in π . We put $R = \hat{\mathbb{Z}}_p$ and again get the commutative diagram (*) with $\hat{\mathbb{Z}}_\pi$ replaced by $\hat{\mathbb{Z}}_p$. We now use Proposition 2 and the embedding $\mathbb{Q}G_1 \to \mathbb{Q}_p G_1$, to conclude that there is a unit $\alpha'' \in \mathbb{Q}_p G_1$ with $\alpha'' G_2 \alpha''^{-1} = G$ and conjugation by α'' induces an automorphism on $\hat{\mathbb{Z}}_p G_1$. But now $\hat{\mathbb{Z}}_p G_1$ is a separable order, and so every automorphism leaving the centre elementwise fixed is conjugation by a unit v'' in $\hat{\mathbb{Z}}_p G$ [8] ; i.e. v'' and α'' act in the same way under conjugation. Since A lies in the radical of $\widetilde{\hat{\mathbb{Z}}_p E_1}$ we argue in the same was as in Case 1. □

THEOREM 7. *Let G be one of the groups in Theorem 2 or a nilpotent class 2 p-group and let E be a finite abelian extension by G . Then E is uniquely determined by its integral group ring, even by the small group ring.*

REMARKS 4. 1) This is the first result on the isomorphism problem which goes much further than circle groups (Sandling [14]) or metabelian groups.

2) Abelian groups are TA for $\mathbb{Z}$ by G. Higman's thesis, so that Theorem 6 also covers metabelian groups.

3. The reduction step

For the sake of simplicity we shall assume that G is a finite p-group, and we take $R = \hat{\mathbb{Z}}_p$. (The results here though hold in more generality.) By induction we assume that a proper image of G has TA, and we shall discuss what is needed to conclude that G has TA.

We first fix some *notation*: Let $S = \hat{\mathbb{Z}}_p[\zeta]$, where ζ is a primitive p-th root of unity over $\hat{\mathbb{Z}}_p$, and put $\tau = (\zeta - 1)$; then $\mathrm{rad}\, S = \tau S$. We now fix a cyclic subgroup $C = \langle c : c^p = 1\rangle$ of order p in the centre of G . Then we have the exact sequence

$$0 \to I(RC).RG \to RG \to RG/C \to 0 .$$

In $\hat{\mathbb{Q}}_p G$ there exists a unique central idempotent ε with $\hat{\mathbb{Q}}_p G\varepsilon = \mathbb{Q}_p G.I(RG)$. We put

$$\Lambda = RG\varepsilon \ .$$

LEMMA 1. *We have the following commutative diagram with exact rows and columns*

$$\begin{array}{ccccccccc}
 & & & & 0 & & 0 & & \\
 & & & & \downarrow & & \downarrow & & \\
 & & & & pRG/C & = & pRG/C & & \\
 & & & & \downarrow & & \downarrow & & \\
0 & \to & I(RC).RG & \to & RG & \to & RG/C & \to & 0 \\
 & & \| & & \downarrow & & \downarrow & & \\
0 & \to & I(RC).RG & \to & \Lambda & \to & \mathbb{F}_p G/C & \to & 0 \\
 & & & & \downarrow & & \downarrow & & \\
 & & & & 0 & & 0 & &
\end{array}$$

Moreover, $I(RC).RG = \tau\Lambda$ *, and* Λ *is an order over* S .

Proof. It should be noted that on Λ , c-1 acts as $\tau = (\zeta-1)$, and so Λ is a $\mathbb{Z}[\tau] = S$-order. Moreover, the middle sequence is split by multiplication with p on RG/C . The rest of the statements is routine calculations. □

The crucial reduction step is

PROPOSITION 3. *Let the notation be as above.*

If α *is a central automorphism of* Λ *such that for every* $\lambda \in \Lambda$:

$$\lambda^{\alpha} \equiv \lambda \bmod \tau \cdot \mathrm{rad}\ \Lambda \ .$$

Then α *is an inner automorphism modified by transfer. Under this assumption* G *is* TA *with respect to* R *by induction.*

Proof. Let ϕ be a normalized automorphism of RG (cf. proof of Theorem 6) then in order to prove TA we have to show that ϕ is inner followed by a group automorphism. By induction hypothesis — cf. the proof of Theorem 6 — we may assume that

$$x^{\phi} \equiv x \text{ modulo } I(RC)I(RG)$$

for every $x \in RG$. Because of Lemma 1, ϕ induces an automorphism α on Λ with

$$\lambda^{\alpha} \equiv \lambda \bmod \tau \cdot \mathrm{rad}\ \Lambda \ ;$$

note $I(RG)\cdot\Lambda = \mathrm{rad}\ \Lambda$. According to the assumption there exists a unit v' in Λ such that for every $\lambda \in \Lambda$ we have

$$\lambda^{\phi} = v' \lambda v'^{-1} \bmod \text{transfer} \ .$$

If we lift v' to a unit v of RG, then in general conjugation by v will not act trivially modulo $\tau\Lambda$. Here one has to do some modifications using Lemma 1 and the fact that RG/C is a group ring, in order to obtain a suitable lifted unit. □

Example 1. The dihedral groups and the quaternion groups.

Let

$$D_{2^n,2} = \{s,\ t : s^{2^n} = t^2 = 1,\ s^t = s^{-1}\}$$

be the dihedral group of order 2^{n+1} and

$$H_{2^n,2} = \{s_0,\ t_0 : s_0^{2^n} = 1,\ s_0^{2^{n-1}} = t_0^2,\ s_0^{t_0} = s_0^{-1}\}$$

be the quaternion group of order 2^{n+1}.

If ξ is a primitive 2^n-th root of unity, we put $\omega_n = \xi_n + \xi_n^{-1}$ and $R_n = \hat{\mathbb{Z}}_2[\omega_n]$, $S_n = \hat{\mathbb{Z}}_2[\xi_n]$.

Then we have the following faithful representations:

$$D_{2^n,2}: \quad s \mapsto \begin{pmatrix} 0 & -1 \\ 1 & \omega_n \end{pmatrix} =: \sigma_n' \qquad t \mapsto \begin{pmatrix} -1 & 0 \\ \omega_n & 1 \end{pmatrix} =: \tau_n'$$

$$H_{2^n,2}: \quad s_0 \mapsto \begin{pmatrix} 0 & -1 \\ -1 & \omega_n \end{pmatrix} =: \sigma_0' = \sigma' \qquad t_0 \mapsto \begin{pmatrix} -i & 0 \\ i\omega_n & i \end{pmatrix} =: \tau_0' = i\tau' .$$

We conjugate these matrices by $\alpha_0 = \begin{pmatrix} 1 & 0 \\ 1 & 1 \end{pmatrix}$, and get

$$\sigma_n = \sigma'^{\alpha_0} = \alpha_0^{-1}\, \sigma_0'\, \alpha_0 = \begin{pmatrix} -1 & -1 \\ 2+\omega_n & 1+\omega_n \end{pmatrix}$$

$$\tau_n = \tau'^{\alpha_0} = \begin{pmatrix} -1 & 0 \\ 2\,\omega_n & 1 \end{pmatrix}$$

$$\sigma_0 = \sigma_0'^{\alpha_0} = \sigma_n\ ; \qquad \tau_0 = \tau_0'^{\alpha_0} = i\tau_n\ .$$

Let $\Lambda_n(D)$ be the order generated by σ, τ over R_n; i,e. the Λ for $\mathbb{Z}D_{2^n,2}$ and $\Lambda_n^i(H)$ the order generated by σ_0, τ_0 over S_n; i.e. the Λ for $\hat{\mathbb{Z}}_2[i]H_{2^n,2}$.

LEMMA 2. (i) $\hat{\mathbb{Z}}_2[i] \otimes_{\hat{\mathbb{Z}}_2} \Lambda_n(D) = \Lambda_n^i(H)$.

(ii) (Miyata-Endo-Sekiguchi [1]) $\Lambda_n(D)$ *satisfies the assumption of Proposition 3.*

Proof. (i) is obvious.

As for (ii), we shall *show more*:

Claim 1. Let O_n be the maximal order in $\hat{\mathbb{Q}}_2[\omega_n]$; i.e. $O_n = \hat{\mathbb{Z}}_2[\omega_n]$, then

$$\Lambda_n(D) = \left\{\begin{pmatrix} a & b \\ \omega_n c & a+\omega_n d\end{pmatrix} , \quad a, b, c, d \in O_n\right\} \quad \text{for } n > 2 .$$

$$\Lambda_2(D) = \left\{\begin{pmatrix} a & b \\ 2c & a+2d\end{pmatrix} , \quad a, b, c, d \in \hat{\mathbb{Z}}_2\right\} .$$

We first treat the case $n = 2$; then $\omega = 0$ and $R_2 = \hat{\mathbb{Z}}_2$. Put

$$x = \sigma_2 + E_2 = \begin{pmatrix} 0 & -1 \\ 2 & 2\end{pmatrix}$$

$$y = \tau_2 + E_2 = \begin{pmatrix} -2 & 0 \\ 2 & 0\end{pmatrix}$$

then

$$x - y = \begin{pmatrix} 2 & -1 \\ 0 & 2\end{pmatrix} = 2E_2 + \begin{pmatrix} 0 & -1 \\ 0 & 0\end{pmatrix} ,$$

and so $z = \begin{pmatrix} 0 & 1 \\ 0 & 0\end{pmatrix} \in \Lambda_2(D)$. Moreover, $zy = \begin{pmatrix} -2 & 0 \\ 0 & 0\end{pmatrix} \in \Lambda_2(D)$, and $y - zy = \begin{pmatrix} 0 & 0 \\ 2 & 0\end{pmatrix}$ $\in \Lambda_2(D)$. Whence $\Lambda_2(D)$ has the desired form, since it obviously can not be larger.

For arbitrary n we shall use induction and note that

$$\omega_n^2 = (\zeta_n + \zeta_n^{-1})^2 = (\zeta_n^2 + \zeta_n^{-2}) + 2 ,$$

i.e.

$$\omega_n^2 = \omega_{n-1} + 2 .$$

If s and t generate $D_{2^n,2}$, then s^2 and t generate $D_{2^{n-1},2}$ as subgroup. So

$$\sigma_n'^2 = \begin{pmatrix} -1 & -\omega_n \\ \omega_n & -1+\omega_n^2\end{pmatrix} .$$

Then

$$\begin{pmatrix} \omega_n^{-1} & 0 \\ 0 & 1\end{pmatrix} \begin{pmatrix} -1 & -\omega_n \\ \omega_n & -1+\omega_n^2\end{pmatrix} \begin{pmatrix} \omega_n & 0 \\ 0 & 1\end{pmatrix} = \begin{pmatrix} -1 & -1 \\ \omega_n^2 & -1+\omega_n^2\end{pmatrix} .$$

Using the relation $\omega_n^2 = \omega_{n-1} + 2$, we find that this matrix is

$$\begin{pmatrix} -1 & -1 \\ \omega_{n-1}+2 & 1+\omega_{n-1} \end{pmatrix} = \sigma_{n-1} .$$

Moreover

$$\begin{pmatrix} \omega_n^{-1} & 0 \\ 0 & 1 \end{pmatrix} \begin{pmatrix} -1 & 0 \\ \omega_n & 1 \end{pmatrix} \begin{pmatrix} \omega_n & 0 \\ 0 & 1 \end{pmatrix} = \begin{pmatrix} -1 & 0 \\ \omega_n^2 & 1 \end{pmatrix} = \begin{pmatrix} -1 & 0 \\ \omega_{n-1}+2 & 1 \end{pmatrix} = \tau_{n-1} .$$

By induction σ_{n-1} and τ_{n-1} generate the order

$$\Lambda_{n-1}(D) = \left\{ \begin{pmatrix} a & b \\ \omega_{n-1}c & a+\omega_{n-1}d \end{pmatrix} : a, b, c, d \in O_{n-1} \right\} \quad \text{for } n > 3$$

or

$$\Lambda_2(D) = \left\{ \begin{pmatrix} a & b \\ 2c & a+2d \end{pmatrix} , a, b, c, d \in \hat{\mathbb{Z}}_2 \right\} \quad \text{for } n = 3 .$$

Thus for $n > 3$, $\Lambda_n(D)$ contains the order

$$\begin{pmatrix} \omega_n & 0 \\ -\omega_n & 1 \end{pmatrix} \begin{pmatrix} a & b \\ \omega_{n-1}c & a+\omega_{n-1}d \end{pmatrix} \begin{pmatrix} \omega_n^{-1} & 0 \\ 1 & 1 \end{pmatrix} = \begin{pmatrix} a+\omega_n b & \omega_n b \\ \omega_n^{-1}\omega_{n-1}c-\omega_n b+\omega_{n-1}d & a-\omega_n b+\omega_{n-1}d \end{pmatrix} ,$$

$a, b, c, d \in O_{n-1}$. In particular, $\Lambda_n(D)$ contains the element

$$\begin{pmatrix} 0 & 0 \\ \omega_n^{-1}\omega_{n-1} & 0 \end{pmatrix} .$$

On the other hand, $\Lambda_n(D)$ has exactly two non-isomorphic irreducible lattices [7] and so the centre of $\Lambda_n(D)$ is maximal; i.e. it is O_n . But

$$\omega_n^{-1}\omega_{n-1}O_n = \omega_n O_n .$$

Thus

$$\begin{pmatrix} 0 & 0 \\ \omega_n & 0 \end{pmatrix} \in \Lambda_n(D) .$$

But then — since

$$\sigma_n - \tau_n = \begin{pmatrix} 0 & -1 \\ 0 & \omega_n \end{pmatrix} ,$$

we have

$$\begin{pmatrix} 1 & 0 \\ \omega_n & 1 \end{pmatrix} \begin{pmatrix} 0 & -1 \\ 1 & \omega_n \end{pmatrix} = \begin{pmatrix} 0 & -1 \\ 0 & 0 \end{pmatrix} \in \Lambda_n(D) .$$

Hence $\Lambda_n(D)$ has the desired form. The proof for $n = 3$ is done similarly.

If now α is conjugation by $a \in \hat{\mathbb{Q}}_p\Lambda_n(D)$ which acts as identity modulo $2\,\mathrm{rad}\Lambda_n(D)$, then α leaves $\mathrm{rad}\Lambda_n(D)$ invariant, and so it acts on the hereditary overorder

$$\Gamma = \begin{pmatrix} O_n & O_n \\ \omega_n O_n & O_n \end{pmatrix}$$

as identity modulo 2Γ ; but then easy calculations show that α is conjugation by $1+\omega_n x$, $x \in \Gamma$ and thus $1+\omega_n x \in \Lambda_n(D)$ is a unit there. Hence $\Lambda_n(D)$ has TA.

For later application we record

LEMMA 3.

$$\Lambda_n^i(H) = \left\{ \begin{pmatrix} a & b \\ \omega_n c & a+\omega_n d \end{pmatrix} , \ a, b, c, d \in S_n \right\} .$$

Proof. This is obvious from the description of $\Lambda_n(D)$.

It should be noted that

$$\hat{\mathbb{Z}}_2[i] \otimes_{\hat{\mathbb{Z}}_2} \Lambda_n(D) = \Lambda_n^i(H) ,$$

so over $\hat{\mathbb{Z}}_2[i]$ the group rings of $D_{2^n,2}$ and $H_{2^n,2}$ almost coincide. They just differ by a map in the pullback of Lemma 1.

4. Connection with piccent $(\hat{\mathbb{Z}}_p G)$

In this section let R be a complete Dedekind domain and Λ and R-order in a separable semisimple finite dimensional algebra over the field of fractions of R . Let S be a finite extension of R which is R-free.

The next is a Noether-Deuring-type Theorem.

PROPOSITION 4. *Let α be a central automorphism of Λ and assume that $1_S \otimes \alpha$ as automorphism of $S \otimes_R \Lambda$ is inner. Then α was inner to start with.*

For the proof we have to invoke some more machinery [2], [3]. Let Pic(Λ) be the isomorphism classes of invertible (Λ, Λ)-bimodules, and Piccent(Λ) be the isomorphism classes of invertible (Λ, Λ)-bimodules X with $zx = xz$ for every $x \in X$, $z \in Z$, the centre of Λ .

We let Autcent(Λ) stand for the central automorphisms of Λ and Inn(Λ) for the inner automorphisms of Λ ; i.e. conjugation with units in Λ . Then

$$\text{Outcent}(\Lambda) = \text{Autcent}(\Lambda)/\text{Inn}(\Lambda)$$

is the group of outer central automorphisms of Λ .

There is a natural map

$$\tilde{\Phi} : \text{Autcent}(\Lambda) \to \text{Piccent}(\Lambda)$$

$$\alpha \quad \mapsto \quad (\Lambda_\alpha)$$

where Λ_α is a (Λ, Λ)-bimodule in the following way: $\lambda_1.x.\lambda_2 = \lambda_1 x(\lambda_2^\alpha)$. It was shown by Fröhlich [2] that Φ induces an *isomorphism*

$$\Phi : \mathrm{Outcent}(\Lambda) \xrightarrow{\sim} \mathrm{Piccent}(\Lambda) .$$

Proof. We use the above machinery. Let $\alpha \in \mathrm{Autcent}(\Lambda)$ and let (α) be its class in $\mathrm{Outcent}(\Lambda)$. Under Φ the class of α corresponds to the isomorphism class of a bimodule X. This isomorphism class is given as the isomorphism class of left $\Lambda\otimes_Z\Lambda^{op}$-modules. The assumption that $1_S\otimes\alpha$ is inner is equivalent to $S\otimes_R X \simeq S\otimes_R\Lambda$ as $S\otimes_R(\Lambda\otimes_Z\Lambda^{op})$-modules. Now we can invoke the ordinary Noether-Deuring theorem for lattices over orders [12] to conclude $X \simeq \Lambda$ as $\Lambda\otimes_Z\Lambda^{op}$-modules; i.e. $(X) = (\Lambda)$ in $\mathrm{Piccent}(\Lambda)$ and consequently $(\alpha) = (1)$ in Outcent-(Λ); hence α is inner. □

COROLLARY 1 (Fröhlich [2]). *The quaternion group of order* 8 *satisfies the assumption of Proposition* 3.

Proof. We have shown in Example 1 that for $\hat{\mathbb{Z}}_2[i]H_{4,2}$ we have the assumption. Now thanks to Proposition 4 we have the assumption for $\hat{\mathbb{Z}}_2 H_{4,2}$. □

REMARK 5. The condition TA says almost, that $\mathrm{Piccent}(\hat{\mathbb{Z}}_p G) = 1$, except for group homomorphisms, which act centrally on $\hat{\mathbb{Z}}_p G$. More precisely: Let $\mathrm{Aut}_C(G)$ be the automorphisms ϕ of G such that g and g^ϕ are conjugate. Then $\mathrm{Inn}(G) \subset \mathrm{Aut}_C(G)$ and the elements in $\mathrm{Aut}_C(G)$ induce central automorphisms of $\hat{\mathbb{Z}}_p G$. It can be shown [2] that

$$\mathrm{Out}_C(G) =: \mathrm{Aut}_C(G)/\mathrm{Inn}\, G \hookrightarrow \mathrm{Piccent}(\hat{\mathbb{Z}}_p G) .$$

In view of the result of §3 the condition TA is equivalent to

$$\mathrm{Out}_C(G) \simeq \mathrm{Piccent}(\hat{\mathbb{Z}}_p G) .$$

It is well-known that in general $\mathrm{Out}_C(G) \neq 1$, and so $\mathrm{Piccent}(\hat{\mathbb{Z}}_p G) \neq 1$.

We shall next prove that group rings of the quaternion groups satisfy TA over $\hat{\mathbb{Z}}$. Thanks to Proposition 4 it is enough to prove the corresponding result for $\hat{\mathbb{Z}}_2[i]$. Before we can do so, we need another general fact.

Let R and Λ be as in the beginning of this section. For an R-order Δ we define

$$M(\Delta) = \{x \in K\Delta : x\mathrm{rad}\Delta + (\mathrm{rad}\Delta)x \subset \mathrm{rad}\Delta\} ,$$

where K is the field of fractions of R. Inductively we put

$$M^i(\Delta) = M(M^{i-1}(\Delta)) , \qquad i > 2$$

$$M^1(\Delta) = M(\Delta) ,$$

and call $M^i(\Delta)$ the *i-th ring of multipliers of* $\mathrm{rad}\Delta$. It is well known that $M(\Delta) \supsetneqq \Delta$ if and only if Δ is not hereditary. Since all our rings are noetherian, there exists a smallest i_o such that

$$M^{i_o}(\Lambda) = \Gamma$$

is hereditary.

LEMMA 4. *Let* α *be a central automorphism of* Λ*; then* $M^\alpha(\Lambda) = M(\Lambda)$ *, i.e.* α *can be extended to the ring of multipliers; in particular* $\Gamma^\alpha = \Gamma$.

Proof. Since $\mathrm{rad}\Lambda$ is the unique maximal ideal, which is nilpotent modulo $\mathrm{rad}R.\Lambda$, we have

$$(\mathrm{rad}\Lambda)^\alpha = \mathrm{rad}\Lambda .$$

Hence if $\mu \in M(\Lambda)$, then μ^α is well-defined since α induces an automorphism of $K\Lambda$, and

$$\mu^\alpha\mathrm{rad}\Lambda = \mu^\alpha(\mathrm{rad}\Lambda)^\alpha = (\mu\mathrm{rad}\Lambda)^\alpha \subset (\mathrm{rad}\Lambda)^\alpha = \mathrm{rad}\Lambda .$$

Similarly $(\mathrm{rad}\Lambda)\mu^\alpha \subset \mathrm{rad}\Lambda$, and so $\mu^\alpha \in M(\Lambda)$; i.e. α induces an automorphism $M(\Lambda)$. □

REMARK 6. If $\alpha \equiv id \bmod \mathrm{rad}R\mathrm{rad}\Lambda$ on Λ , then $\alpha \equiv id \bmod \mathrm{rad}R.M(\Lambda)$ on $M(\Lambda)$, as is easily seen.

PROPOSITION 5. *Let* $H_{2^n,2}$ *be the quaternion group of order* 2^{n+1} . *Then* $\hat{\mathbb{Z}}_2H_{2^n,2}$ *satisfies* TA.

Proof. In view of Proposition 3 and with the notation of §3 as developed for the quaternion groups (Lemma 3), we have to show the following — having applied already Proposition 4:

Let α be a central automorphism of $\Lambda_n^i(H)$ with $\alpha \equiv id \bmod 2\mathrm{rad}\Lambda_n^i(H)$. Then α is inner.

Let $\pi = (1-\xi_n)$ be a parameter of $S_n = \hat{\mathbb{Z}}_2[\xi_n]$, then $\Lambda_n^i(H)$ can be written in the form

$$\Lambda_n^i(H) = \left\{\begin{pmatrix} a & b \\ \pi^2 c & a+\pi^2 d\end{pmatrix} , \ a, b, c, d \in S_n\right\} .$$

If we put — for the time being —

$$\Lambda = \left\{\begin{pmatrix} a & b \\ \pi c & a+\pi^2 d\end{pmatrix} , \ a, b, c, c \in S_n\right\}$$

then Λ is conjugate to $\Lambda_n^i(H)$ and so α induces an automorphism of Λ which is trivial modulo $2\mathrm{rad}\Lambda$.

We next compute $M(\Lambda) = M^1(\Lambda)$ as

$$M^1(\Lambda) = \left\{\begin{pmatrix} a & b \\ \pi c & a+\pi d \end{pmatrix} , \quad a, b, c, d \in S_n\right\}$$

and

$$M^2(\Lambda) = \left\{\begin{pmatrix} a & b \\ c & d \end{pmatrix} , \quad a, b, c, d \in S_n\right\} .$$

Because of Lemma 4, α acts on $M^2(\Lambda)$ and so α is a unit in $M^2(\Lambda)$ which by Remark 6 acts via conjugation on $M^1(\Lambda)$ trivially modulo 2 . Since $2S_n = \pi^\nu S_n$ with $\nu \geq 2$ one finds easily that α must be conjugation with a unit in Λ . □

Proof of Theorem 5. Let ϕ be a normalized automorphism of $\mathbb{Z}_\pi G$. Let

$$C = \prod_{i=1}^{n} C_i ,$$

C_i cyclic of order p_i in the centre of P_i . By induction, G/C has TA for $\mathbb{Z}_\pi$. As in the proof of Theorem 6 — noting that one can lift units from $\mathbb{Z}_\pi G/C$ to $\mathbb{Z}_\pi G$, we may assume that the automorphism ϕ is trivial modulo

$$2\mathbb{Z}_\pi\, c{\uparrow}^G$$

and so — in particular ϕ is central.* Let ϕ correspond to the bimodule M in $\mathrm{Piccent}(\mathbb{Z}_\pi G)$.

Claim. M is trivial in $\mathrm{Piccent}(\mathbb{Z}_\pi G)$.

Proof. Since M is left module over $\mathbb{Z}_\pi G \otimes_{\mathbb{Z}_\pi} \mathbb{Z}_\pi G^{op}$, it is enough to show that $M_p = \mathbb{Z}_p \otimes M$ is trivial in $\mathrm{Piccent}(\mathbb{Z}_p G)$ for every $p \in \pi$. If $p \neq p_i$ for some i , this is clear. But

$$\hat{\mathbb{Z}}_{p_i} G = \hat{\mathbb{Z}}_{p_i} P_i \otimes_{\hat{\mathbb{Z}}_{p_i}} \hat{\mathbb{Z}}_{p_i} \Big(\prod_{j \neq i} P_j \Big) .$$

Since $\hat{\mathbb{Z}}_{p_i} \big(\prod_{j \neq i} P_j \big)$ is a separable order, it is Morita equivalent to πR_k , where R_k are unramified extensions of $\hat{\mathbb{Z}}_p$, and so $\hat{\mathbb{Z}}_{p_i} G$ is Morita equivalent to $\prod_k R_k P_i$. Under the isomorphism [2]

$$\mathrm{Piccent}(\hat{\mathbb{Z}}_{p_i} G) \simeq \coprod \mathrm{Piccent}(R_k P_i) M_{p_i}$$

and whence $1_{\hat{\mathbb{Z}}_{p_i}} \otimes \phi$ corresponds to a family ϕ_k of central automorphisms of $R_k P_i$ which is trivial modulo $R_k \otimes_{\hat{\mathbb{Z}}_{p_i}} c_i{\uparrow}^{P_i}$. According to our assumption (cf. proof of Theorem 6) ϕ_k is an inner automorphism; i.e. $\hat{M}_{p_i}$ is trivial in $\mathrm{Piccent}(\hat{\mathbb{Z}}_{p_i} G)$. This proves the claim and also the theorem. □

---

* May 1984: This is in fact not true, whence Theorem 5 is false.

5. Non-abelian cohomology, exponentials and logarithms

We return to the notation as introduced in the beginning of §3. In particular, ζ is a primitive p-th root of unity and $\hat{S} = \hat{\mathbb{Z}}_p[\zeta]$, $\tau = (\zeta - 1)$.

We consider cocycles

$$\theta : G \to U(\Lambda) \ , \quad \text{the units in } \Lambda \ ;$$

i.e. maps: $\theta : G \to U(\Lambda)$ with $\theta(gh) = h^{-1}\theta(g)h\theta(h)$ — note that G acts by conjugation on $U(\Lambda)$. Special cocycles are the coboundaries: Let $u \in U(\Lambda)$ and define

$$\begin{aligned} \theta_u : G &\to U(\Lambda) \\ g &\to g^{-1}u^{-1}gu \ . \end{aligned}$$

Then θ_u is a cocycle. Let $H^1(G, (\Lambda))$ be the quotient of the set of cocycles modulo the equivalence relation induced by the coboundaries. Then $H^1(G, U(\Lambda))$ is a pointed set, the *first cohomology* set. Similarly $H^1(G, U)$ is defined for any G-invariant subgroup U of $U(\Lambda)$.

LEMMA 5. *We have an injective map*

$$\begin{aligned} \mathrm{Aut}(\Lambda)/\mathrm{Inn}(\Lambda) &\to H^1(G, U(\Lambda)) \\ \alpha &\mapsto \theta_\alpha : g \to g^{-1}g^\alpha \ , \end{aligned}$$

where we have identified G *with its image in* Λ .

Proof. Obviously θ_α is a cocycle and inner automorphisms go to coboundaries. It is also clear that the map is injective.

Let $H^1(G, U(\Lambda))^*$ be the image under the map in Lemma 5. Note that $H^1(G, U(\Lambda))^*$ can also be defined internally, since every cocycle $\theta : G \to U(\Lambda)$ induces a ring homomorphism

$$\begin{aligned} \phi_\theta : \mathbb{Z}G &\to \Lambda \qquad \text{induced by} \\ g &\to g\theta(g) \ . \end{aligned}$$

In particular, *the assumption of Proposition 3 would be satisfied if we could show* that the induced map

$$(*) \qquad H^1(G, 1 + \tau\mathrm{rad}\Lambda)^* \to H^1(G, 1 + \mathrm{rad}\Lambda)^*$$

is trivial.

Before we come to positive results, let us look for some arguments, why (*) should hold. In analogy to Lie theory, where one gets automorphisms from additive derivations, let us look at the *additive analogue of* (*). We consider $\tau\mathrm{rad}\Lambda$ and $\mathrm{rad}\Lambda$ as abelian groups with G acting via conjugation. Then the additive analogue of (*) is true and follows.

PROPOSITION 6. *The induced map*

$$H^1(G, \tau\mathrm{rad}\Lambda) \to H^1(G, \mathrm{rad}\Lambda)$$

of abelian cohomology groups is trivial.

The proof of this and the following results will be published elsewhere.

In this connection we would like to point out a result on the cyclic group C_p of order p, which is more general than (*), but less sharp for $p > 2$.

PROPOSITION 7. *Let* Λ *be a* $\mathbb{Z}_p$*-order on which* C_p *acts. Then the map*

$$H^1(C_p, 1 + p\mathrm{rad}\Lambda) \to H^1(C_p, 1 + \mathrm{rad}\Lambda)$$

is trivial.

The next result has been very useful to prove TA for special p-groups as considered by Sekiguchi, and it also should be of help in studying wreath products.

PROPOSITION 8. *Let* α *be a central automorphism of* Λ *which is conjugation by* $1 + \tau\delta$, *where* $\delta \in \mathrm{rad}\Gamma$ *for some order* $\Gamma \supset \Lambda$. *If* α *acts as identity modulo* $\tau\mathrm{rad}\Lambda$, *then* α *is inner.*

REMARKS 7. (i) By passing to a larger ground ring — thanks to the Noether-Deuring theorem this is no loss of generality — one can always assume that α is conjugation by $1 + \delta$, δ integral; i.e. $\delta \in \Gamma$ for some order Γ.

(ii) The proof is based on manipulation with *exponentials* and *logarithms*. In fact we have

1) $\exp(\tau\delta)$ (barely) converges

2) $\log(\exp(\tau\delta))$ converges and is equal to $\tau\delta$.

Further manipulations with exponentials and logarithms allow to almost verify (*); in fact we have

THEOREM 8. *Let* α *be a central automorphism of* Λ *which acts trivially modulo* $\tau\mathrm{rad}^2\Lambda$, *then* α *is inner. Equivalently the map*

$$H^1(G, 1 + \tau\mathrm{rad}^2\Lambda)^* \to H^1(G, 1 + \mathrm{rad}\Lambda)^*$$

is trivial.

This theorem is the crucial ingredient to the proofs of Theorems 3 and 4, since in that case one can pass from $\tau\mathrm{rad}\Lambda$ to $\tau\mathrm{rad}^2\Lambda$.

REMARK 8. Theorem 8 holds for more general orders Δ (instead of Λ), provided $\Delta = R + \mathrm{rad}\Delta$ and τ-multiples of R-derivations are inner.

References

[1] S. Endo, T. Miyata, and K. Sekiguchi, "Picard groups and automorphism groups of dihedral group rings of metacyclic groups", Preprint.

[2] A. Fröhlich, "The Picard group of non-commutative rings, in particular of orders", *Trans. Amer. Math. Soc.* 180 (1973), pp. 1-45, MR47:6751.

[3] A. Fröhlich, I. Reiner, and S. Ullom, "Class groups of orders and Picard groups", *Proc. London Math. Soc.* (3) 29 (1974), pp. 405-434, MR50:9932.

[4] K.W. Gruenberg and K.W. Roggenkamp, "Extension categories of groups and modules, I, Essential covers", *Journal of Algebra* 49 (1977), pp. 564-594, MR58:16780
"Extension categories of groups and modules, II, Stem extensions", *Journal of Algebra* 67 (1980), pp. 342-368, MR82i:16031.

[5] G. Higman, "Units in group rings", D.Phil. Thesis, University of Oxford (1940).

[6] D.A. Jackson, "The groups of units of the internal group rings of finite metabelian and finite nilpotent groups", *Quart. J. Math. Oxford* (2) 20 (1969), pp. 319-331, MR40:2766.

[7] C.R. Leedham-Green and M.F. Newman, "Space groups and groups of prime power order I", *Archiv Math.* 35 (1980), pp. 193-202, MR81m:20029.

[8] I. Reiner, *Maximal Orders* (London Math. Soc. Lecture Notes 5, Academic Press, 1975), MR52:13910.

[9] I. Reiner and K.W. Roggenkamp, *Integral Representations* (Lecture Notes in Mathematics 744, Springer-Verlag, 1979), pp. 1-143, 149-275, MR80k:20010.

[10] I. Reiner and H. Zassenhaus, "Equivalence of representations under extensions of local ground rings", *Ill. J. Math.* 5 (1961), pp. 409-411, MR23:A3764.

[11] J. Ritter and S.K. Sehgal, "On a conjecture of Zassenhaus on torsion units in integral group rings", Preprint, Augsburg (1983).

[12] K.W. Roggenkamp, "Group rings of metabelian groups and extension categories", *Canad. J. Math.* 32 (1979), pp. 449-459, MR81f:16019.

[13] K.W. Roggenkamp, "Units in integral metabelian group rings, I. Jackson's unit theorem revisited", *Quart. J. Math. Oxford* (2) 32 (1981), pp. 209-224, MR82i:16009.

[14] R. Sandling, "Group rings of circle and unit groups", *Math. Zeitschrift* 140 (1974), pp. 195-202, MR52:3217.

[15] K. Sekiguchi, "On the automorphism group of the p-adic group ring of a metacyclic p-group", Preprint.

[16] A. Whitcomb, "The group ring problem", Ph.D Thesis, University of Chicago (1968).

[17] H. Zassenhaus, "On the torsion units of finite group rings", *Studies in Mathematics (in honor of A. Almeida Costa)*, Instituto de Alta Cultura, Lisboa (1974), pp. 119-126 (Portuguese), MR51:12922.

Mathematisches Institut B,
Universität Stuttgart,
7000 Stuttgart-80, Federal Republic of Germany

PROCEEDINGS OF 'GROUPS — KOREA 1983' 20C07
KYOUNGJU, August 1983

ON INDUCED ISOMORPHISMS OF GROUP RINGS

Frank Röhl

1. Introduction

If G and H are groups and $\mathbb{Z}G$ resp. $\mathbb{Z}H$ their integral group rings, one might suspect that a "well behaved" isomorphism $\psi : \mathbb{Z}G \to \mathbb{Z}H$ should be induced by an isomorphism of the groups. But this seems to be a rare phenomenon: The only finite groups for which all isomorphisms are induced, are abelian or hamiltonian 2-groups (see section 2). (Without loss of generality we may and will assume that every isomorphism of group rings is augmented.)

So one has to look for the least worse situation: Not the isomorphism ψ itself is induced but modulo a suitable ideal ψ is induced by an isomorphism of the groups, i.e.

$$\begin{array}{ccc} G & \xrightarrow{\sim} & H \\ \downarrow & & \downarrow \\ \mathbb{Z}G/I & \longrightarrow & \mathbb{Z}H/\psi I \end{array} .$$

For brevity let us call ψ to be "induced modulo I". Nearly all results on the isomorphism problem for integral group rings may be viewed from this position: If G is the circle group of a nilpotent ring, ψ is induced modulo the complement of $\delta(G)$ (where $\delta(g) := g-1$ is a monomorphism of G into the augmentation ideal ΔG of $\mathbb{Z}G$) in ΔG (see [3] and [4]), and another result in this direction is the following theorem due to Whitcomb (see for example [5], Thm. III.5.3 on p.102): If G is a torsion-group and $\psi : \mathbb{Z}G \xrightarrow{\sim} \mathbb{Z}H$, then $G/G'' \xrightarrow{\sim} H/H''$, where $G' := [G, G]$ and $G'' := [G', G']$, $[g_1, g_2] := g_1^{-1}g_2^{-1}g_1g_2$ for $g_1, g_2 \in G$. It turns out that ψ is induced modulo $\Delta G \Delta(G, G')$, $\Delta(G, G')$ denoting the ideal generated by all $g-1$, $g \in G'$.

Analyzing the proof of this theorem, it is easy to give the essence of it and thus to obtain that nilpotent groups G with abelian torsion-subgroup are characterized by $\mathbb{Z}G$ (see section 4).

As a technical tool on this way, we show that if G is nilpotent and $\mathbb{Z}G \xrightarrow{\sim} \mathbb{Z}H$, then H , too, is nilpotent (see section 3), giving thereby an answer to a question raised by Sehgal (see Problem 25 on p.229 in [5]). The technique used in the proof also yields a very simple proof of the fact that the class of nilpotence of a torsion-group is preserved under isomorphisms of integral group rings.

In the last section we start an investigation whether the concept of induced isomorphisms also works in modular group rings. Since the analogue of Whitcomb's theorem in this case is already established by Passi and Sehgal in [1], we only try to translate Sandling's result on finite circle groups. The difficulty here is that we have no correspondence of ideals "disjoint" from G nor triviality of central torsion-units. We obtain that the circle group G of a nilpotent, finite dimensional F_p-algebra A, F_p the field of p elements, is characterized by F_pG, if the Frattini-subgroup of G coincides with A^2.

2. Induced isomorphisms

Obviously, all isomorphisms $\psi : \mathbb{Z}G \to \mathbb{Z}H$ are induced by an isomorphism of the groups, if the units in $\mathbb{Z}G$ are trivial. This is the case, for example, for torsion-free nilpotent groups. One may ask now, whether the above condition is necessary. An answer, at least if G is finite, is given by

PROPOSITION. *Let G be a finite group. Every isomorphism $\mathbb{Z}G \tilde{\to} \mathbb{Z}H$ is induced, iff the torsion-units of $\mathbb{Z}G$ are trivial (and this happens exactly, if G is abelian or a hamiltonian 2-group).*

Proof. Let H be a subgroup of the group $U_1(\mathbb{Z}G)$ of units of augmentation 1 of $\mathbb{Z}G$. The inclusion $H \hookrightarrow U_1(\mathbb{Z}G)$ induces a homomorphism $\mathbb{Z}H \to \mathbb{Z}G$, which is an isomorphism in case H is a $\mathbb{Z}$-base of $\mathbb{Z}G$. By our assumption, this isomorphism is induced; hence, $H = G$! In particular, G is stable under every automorphism of $\mathbb{Z}G$ given by conjugation with a unit. Thus, G is normal in $U_1\,\mathbb{Z}G$, and by a theorem of Berman, Sehgal, and Zassenhaus (see Thm. II.2.18 on p.53 in 5) the torsion-units of $\mathbb{Z}G$ are trivial, and G is of the stated type by the same theorem.

The other direction is clear. □

Note that there are finite abelian groups G such that $\mathbb{Z}G$ possesses non-trivial units!

In any case, the above proof shows that G is the only subgroup of $U_1(\mathbb{Z}G)$, which is a $\mathbb{Z}$-base of $\mathbb{Z}G$ and hence, G is normal in $U_1(\mathbb{Z}G)$, if every isomorphism $\mathbb{Z}G \tilde{\to} \mathbb{Z}H$ is induced.

3. The torsion-subgroup correspondence

If G is any group, let $T(G)$ denote the set of all torsion-elements of G, and write $T_i(G) := T(G) \cap Z_i(G)$, $i \geq 0$, where $Z_i(G)$ are the terms of the upper central-series of G.

If G is nilpotent, the $T_i(G)$ form an ascending chain of normal subgroups with the property

$$T_i(G/T_j(G)) = T_{i+j}(G)/T_j(G) .$$

Let the "torsion-length $t(G)$ " of G be the number of different terms $T_i(G)$ $(\neq 1)$, then $t(G) = 0$ implies that G is torsion-free (or 1), because every normal subgroup $(\neq 1)$ of G meets the center non-trivially. For $t(G) > 0$, we obtain from the above formula

$$t(G) = t(G/T_1(G)) + 1 .$$

PROPOSITION. *If G is a nilpotent group and H a group such that $\mathbb{Z}G \cong \mathbb{Z}H$, then H is nilpotent and $t(G) = t(H)$.*

Proof. We proceed by induction on the torsion-length of G:

If $t(G) = 0$, then G is torsion-free and the units of $\mathbb{Z}G$ are trivial. Hence, $G \cong H$.

Let $t(G) > 0$. Now, central units of finite order are trivial, so, if we identify $\mathbb{Z}G$ and $\mathbb{Z}H$, this gives $T_1(G) \subset T_1(H)$ and by symmetry, $T_1(G) = T_1(H)$.

The sequence $1 \to T_1(H) \to H \to H/T_1(H) \to 1$ is exact and by the natural isomorphism $\mathbb{Z}H/\Delta(H, T_1(H)) \cong \mathbb{Z}(H/T_1(H))$ and inductive hypothesis, $H/T_1(H)$ is nilpotent of torsion-length $t(G/T_1(G))$. Since $T_1(H)$ is central, the assertion follows. □

From this proposition and triviality of central torsion-units follows immediately the first part of the next

LEMMA. *Let G be a nilpotent group and H a group with $\mathbb{Z}G = \mathbb{Z}H$. Then one has*

a) $\Delta(G, T(G)) = \Delta(H, T(H))$

b) *If $I \subset \Delta(G, T(G))$ is an ideal with $(I+1) \cap G = 1$, then $(I+1) \cap H = 1$.*

Proof. Assume $(I+1) \cap H \neq 1$. Then, since H is nilpotent and $(I+1) \cap H$ is a normal subgroup, $(I+1) \cap H$ meets the center $Z_1(H)$ of H non-trivial. Since $I+1$ does not meet G and by triviality of central torsion-units, $(I+1) \cap Z_1(H) \neq 1$ is torsion-free. Thus, $H/T(H) \to \Delta H/\Delta(H, T(H)) \cong \Delta(H/T(H))$, where the first map is given by $h \mapsto h-1+\Delta(H, T(H))$, is not injective, a contradiction. □

REMARK. If G is a nilpotent torsion-group, the class of G coincides with the torsion-length and from a) of the foregoing lemma together with the proposition it follows that in this case the class of nilpotence is preserved under integral group ring isomorphisms. More general: If G is a group such that $T(G)$ is a subgroup and H is a group with $\mathbb{Z}G = \mathbb{Z}H$, then the map

$$T_i(G) \mapsto H \cap (1+\Delta(G, T_i(G)))$$

and vice versa defines a bijective correspondence $T_i(G) \leftrightarrow T_i(H)$.

4. Remarks on Whitcomb's theorem

We collect the necessary ingredients of Whitcomb's theorem in the following

LEMMA. *Let* G *and* H *be groups with* $\psi : \mathbb{Z}G \xrightarrow{\sim} \mathbb{Z}H$ *, and let* N *be a normal subgroup of* G *with the following properties:*

(i) *There exists a normal subgroup* N_H *of* H *such that* $\psi\Delta(G, N) = \Delta(H, N_H)$ *and* ψ *is induced* modulo $\Delta(G, N)$ *by an isomorphism* $G/N \xrightarrow{\sim} H/N_H$.

(ii) *There exists an ideal* I *in* $\mathbb{Z}G$ *and an isomorphism* $N \xrightarrow{\sim} N_H$ *with*

$$\begin{array}{ccc} N & \xrightarrow{\sim} & N_H \\ \bar{\delta} \downarrow & & \downarrow \bar{\delta} \\ \Delta(G, N)/I & \xrightarrow{\sim} & \Delta(H, N_H)/\psi I \end{array}$$

where $\bar{\delta}$ *, defined by* $\bar{\delta}(n) := n - 1 + I$ *, is bijective.*

Then ψ *is induced* modulo I .

Sketch of Proof. Let us identify $\mathbb{Z}G$ and $\mathbb{Z}H$ and define δ by $\delta(n) := n - 1$ for n a group element. Then, for $h \in H$, there is a $g \in G$ and an $x \in \Delta(G, N)$ with $\delta(h) = \delta(g) + x$ by (i). With $y := g^{-1}x \in \Delta(G, N)$ one obtains $\delta(h) = \delta(g) + y + \delta(g)y =: \delta(g) \circ y$.

By (ii) there exists a $g' \in N$ with $y = \delta(g') + y'$, $y' \in I$.

Hence, $\delta(H) + I \subset \delta(G) + I$, and equality holds by symmetry. In view of (ii), one has $\delta(H) \cap I = 0 = \delta(G) \cap I$. □

Whitcomb's theorem is an easy consequence of the foregoing: Take $N := G'$. Since $\Delta(G, N)$ is the minimal ideal such that $\mathbb{Z}G/\Delta(G, N)$ is commutative, and since G/N is an abelian torsion-group, (i) holds; and (ii) follows with $I = \Delta G\Delta(G, N)$ (see [5], Prop. III.4.20 and Thm. III.4.28). Thus, for torsion-groups G , G/G'' is characterized by $\mathbb{Z}G$.

But we can do a little more.

COROLLARY. *Let* G *be a nilpotent group such that the torsion* $T(G)$ *is abelian. Then* G *is characterized by* $\mathbb{Z}G$*. (Since the torsion-subgroup of a divisible nilpotent group is central by Chernikov's theorem, these groups are included in this class of groups.)*

Proof. We show that every isomorphism $\mathbb{Z}G \xrightarrow{\sim} \mathbb{Z}H$ is induced modulo $\Delta G\Delta(G, T(G))$.

By part a) of the lemma in section 3 we know that (i) holds, since the units of $\mathbb{Z}(G/T(G))$ are trivial; and by proposition III.1.15 on p.76 in [5], there is an isomorphism $T(G) \to \Delta(G, T(G))/\Delta G\Delta(G, T(G))$ given by $n \mapsto n - 1 + \Delta G\Delta(G, T(G))$. Now (ii) follows in view of part b) of the lemma in section 3. □

5. The modular case

For a finite p-group G, p a prime, let $M_2(G)$ denote its second dimension-subgroup with respect to F_p. Then it is well-known that $M_2(G)$ is just the Frattini-subgroup of G. Now, we can prove

THEOREM. *Let G be the circle group of a nilpotent, finite dimensional F_p-algebra A. If $M_2(G) = A^2$, then G is characterized by F_pG.*

Proof. By the universal property of the augmentation ideal Δ_pG of F_pG with respect to circle groups, one has

$$\begin{array}{ccc} G & \xrightarrow{Id} & A \\ \delta\downarrow & \nearrow_{\phi} & \\ \Delta_pG & & \end{array} \quad , \ \delta(g) = g-1 \ , \ \phi \text{ an } F_p\text{-algebra-homomorphism.}$$

We show that every isomorphism $\psi : F_pH \xrightarrow{\sim} F_pG$ is induced modulo $I := \ker \phi$.

First of all, there is a homomorphism $\phi\psi\delta_H : H \to G$ (with $\delta_H(h) = h-1$), and by $|G| = |H| = \dim F_pG < \infty$, it is sufficient to show that $\phi\psi\delta_H$ is surjective.

Now, the above diagram induces

$$\begin{array}{ccc} G/M_2(G) & \longrightarrow & A/A^2 \\ \bar{\delta}\downarrow & \nearrow_{\bar{\phi}} & \\ \Delta_pG/(\Delta_pG)^2 & & \end{array} \quad , \ \text{where } \bar{\delta} \text{ is an isomorphism,}$$

and an easy diagram chase shows the condition $M_2(G) = A^2$ to be equivalent to $\bar{\delta}^{-1} = \bar{\phi}$ (and to $I \subset (\Delta_pG)^2$).

Hence, the map $(\phi\psi\delta_H)_2 : H/M_2(H) \to G/M_2(G)$ induced by $\phi\psi\delta_H$ coincides with the following map

$$H/M_2(H) \xrightarrow{\sim} \Delta_pH/(\Delta_pH)^2 \xrightarrow{\sim} \Delta_pG/(\Delta_pG)^2 \xrightarrow{\sim} G/M_2(G)\,,$$

which is induced by $\bar{\delta}^{-1}\psi\delta_H$ and thus is surjective.

Now, since the Frattini-subgroup of G (resp. H) is just the set of all non-generators of G (resp. H), the original homomorphism $\phi\psi\delta_H$ must be already surjective. □

REMARKS. 1. The author would like to thank Robert Sandling, who told him the elegant Frattini-subgroup-argument of the last part of the proof, thereby avoiding the clumsiness of an earlier version!

2. Another result on the isomorphism problem for circle groups of nilpotent F_p-

algebras is contained in [2]. It seems that the lack of the "disjoint ideal correspondence" resp. triviality of central torsion-units forces one to impose rather severe restrictions on the algebras under consideration: We assumed $M_2(G) = A^2$, where Roggenkamp and Scott assumed $A^p = 0$ and $Z(A)^2 = 0$, $Z(A)$ the centre of A.

Nevertheless, as an application of our theorem, we obtain a beautiful remark on the group of upper triangular matrices with coefficients in F_p. Since it is the circle group of the ring of strictly upper triangular matrices (zeros on and below the diagonal), one easily verifies that it satisfies the hypothesis of the theorem and thus is characterized by its modular group ring.

There is another special case of the theorem, a result due to Sehgal (see [5], Cor. III.6.25 on p.117):

COROLLARY. *If G and H are finite groups with $M_3(G) = 1 = M_3(H)$, $M_3 G$ (resp. $M_3(H)$) denoting the third dimension-subgroup of G (resp. H) with respect to F_p for some prime p, then $F_pG \cong F_pH$ implies $G \cong H$.*

Proof. These groups are circle groups, because $M_2(G)$ admits a vector space complement in $(\Delta_pG)^2/(\Delta_pG)^3$, which obviously can be lifted to an ideal I in Δ_pG. And I is in view of the isomorphism $G/M_2(G) \cong \Delta_pG/(\Delta_pG)^2$ a complement of $\delta(G)$ in Δ_pG with $I \subset (\Delta_pG)^2$ (see the proof of the theorem). □

References

[1] I.B.S. Passi and S.K. Sehgal, "Isomorphism of modular group algebras", *Math. Z.* 129 (1972), pp. 65-73, MR47:314.

[2] K.W. Roggenkamp and L. Scott, "The isomorphism problem and units in group rings of finite order", *Groups – St. Andrews 1981* (London Math. Soc. Lecture Notes 71, 1982), pp. 313-327, Zbl.492:16017.

[3] F. Röhl, "On the isomorphism problem for integral group-rings of circle-groups", *Math. Z.* 180 (1982), pp. 409-422.

[4] R. Sandling, "Group rings of circle and unit groups", *Math. Z.* 140 (1974), pp. 195-202, MR52:3217.

[5] S.K. Sehgal, *Topics in Group Rings* (Dekker, New York, 1978), MR80j:16001

Mathematisches Seminar d. Universität Hamburg,
Bundesstr. 55,
2000 Hamburg 13, Federal Republic of Germany

PROCEEDINGS OF 'GROUPS — KOREA 1983'
KYOUNGJU, August 1983 20E06

ÜBER DARSTELLUNGEN VON ELEMENTEN UND UNTERGRUPPEN IN FREIEN PRODUKTEN

GERHARD ROSENBERGER

Einleitung

Sei $G = H_1 * \ldots * H_n$ das freie Produkt der Gruppen $H_1, \ldots, H_n$. Sei $1 \neq a_j \in H_j$ und p die Anzahl der a_j, die echte Potenz in H_j sind $(1 \leq j \leq n)$.

Sei $\{x_1, \ldots, x_m\} \subset G$ $(m \geq 1)$ und H die von $x_1, \ldots, x_m$ erzeugte Untergruppe von G.

Ist $a_1 \ldots a_n \in H$, so tritt einer der folgenden Fälle ein (Satz (2.1)):

(1) Es gibt einen freien Übergang von $\{x_1, \ldots, x_m\}$ zu einem System $\{y_1, \ldots, y_m\}$ mit $y_1 = a_1 \ldots a_n$;

(2) es ist $m \geq 2n-p$, und es gibt einen freien Übergang von $\{x_1, \ldots, x_m\}$ zu einem System $\{y_1, \ldots, y_m\}$ mit $y_i \in H_j$ $(1 \leq j \leq n,$ $1 \leq i \leq 2n-p)$.

Dieser Satz besagt also, dass sich H in geeigneter Weise durch $a_1 \ldots a_n$ beschreiben lässt. Er wurde bereits in [15] formuliert, allerdings wurde dort auf einen Beweis verzichtet, da sich sonst gewisse Wiederholungen in [15] ergeben hätten. Nun wurden in letzter Zeit mehrere Resultate über Darstellbarkeit von Elementen in freien Produkten (und speziell in freien Gruppen) veröffentlicht, die zum grossen Teil lediglich direkte Folgerungen aus Satz (2.1) oder Spezialfälle von Satz (2.1) sind (vergleiche zum Beispiel [2] und einige der dort zitierten Arbeiten). Daher erscheint uns jetzt eine Veröffentlichung des Beweises von Satz (2.1) doch noch angebracht, um etwas stärker auf diesen Satz hinzuweisen.

Wir geben dann einige Folgerungen aus Satz (2.1) und fragen schliesslich nach möglichen Erweiterungen für den Fall, dass nicht $a_1 \ldots a_n$ in H, wohl aber eine Potenz $(a_1 \ldots a_n)^\alpha$, $\alpha > 0$, in H liegt. Bemerkenswert scheint uns dabei besonders der folgende Satz zu sein, durch den die Ergebnisse aus [5] und [18] ergänzt werden (Satz (2.15)):

Sei $G = \langle s_1, \ldots, s_n \mid s_1^{\alpha_1} = \ldots = s_n^{\alpha_n} = 1 \rangle$, $2 \le n$, die α_i Primzahlen ($i = 1, \ldots, n$) , und es sei $\alpha_1 + \alpha_2 \ge 5$ falls $n = 2$ ist (G sei also nicht isomorph zur unendlichen Diedergruppe). Seien $x_1, \ldots, x_m$ ($m \ge 1$) Elemente endlicher Ordnung in G , und sei H die von $x_1, \ldots, x_m$ erzeugte Untergruppe von G .

Genau dann hat H in G endlichen Index, wenn $y(s_1 \ldots s_n)^{\alpha} y^{-1} \in H$ für ein $\alpha \ne 0$ und ein $y \in G$.

Darüberhinaus gilt: Hat H in G den endlichen Index $[G : H]$ und ist α die kleinste positive Zahl mit $y(s_1 \ldots s_n)^{\alpha} y^{-1} \in H$ für ein $y \in G$, so ist $\alpha = [G : H]$, und ein Repräsentantensystem für die Rechtsrestklassen von G nach H ist gegeben durch $\{1, s_1 \ldots s_n, \ldots, (s_1 \ldots s_n)^{\alpha - 1}\}$.

1. Vorbemerkungen

A. Diese Arbeit verwendet die Terminologie und Bezeichnungsweise von [5], [15] und [17], wobei $\langle \ldots \mid \ldots \rangle$ die Gruppenbeschreibung durch Erzeugende und Relationen bedeutet. Die verwendeten Begriffe und Sätze aus der kombinatorischen Gruppentheorie finden sich in [8] und [19].

Es bedeute:

$\langle x_1, \ldots, x_m \rangle$ die von $x_1, \ldots, x_m$ erzeugte Gruppe;

$Z_\alpha := \langle s \mid s^\alpha = 1 \rangle$, $\alpha \ge 2$, die zyklische Gruppe der Ordnung α ;

$[x, y] := xyx^{-1}y^{-1}$ den Kommutator von $x, y \in G$ (G Gruppe);

$[G : H]$ den Index von H in G ($H \subset G$ Gruppen);

$(\alpha_1, \ldots, \alpha_m)$ den grössten gemeinsamen Teiler der Zahlen $\alpha_1, \ldots, \alpha_m \in \mathbb{Z}$.

Alle auftretenden Exponenten von Gruppenelementen seien ganze Zahlen.

Sei $G = H_1 * \ldots * H_n$ ($n \ge 1$) das freie Produkt der Gruppen $H_1, \ldots, H_n$.

Wir vereinbaren, dass jede solche Zerlegung nicht-trivial (d.h. stets $H_i \ne \{1\}$ für $i = 1, \ldots, n$) ist.

Jedes $x \in G$ besitzt eine eindeutig bestimmte Darstellung $x = h_1 \ldots h_p$, $p \ge 0$, wobei die h_i abwechselnd in verschiedenen Faktoren H_j von G liegen und ungleich 1 sind.

Durch $L(x) = p$ wird eine Länge von x definiert.

Jedes $x \in G$ besitzt nun die eindeutig bestimmte symmetrische Normalform $x = l_1 \ldots l_m k r_m \ldots r_1$, wobei gilt:

(a) $m \geq 0$;

(b) die l_j bzw. r_j liegen abwechselnd in verschiedenen Faktoren;

(c) für $L(x) = 0$ ist $m = 0$ und $k = 1$;

(d) für $L(x) = 2m$ $(m \geq 1)$ ist $k = 1$, und l_m und r_m liegen in verschiedenen Faktoren von G ; und

(e) für $L(x) = 2m + 1$ ist $k \neq 1$, $k \in H_i$ für einen Faktor H_i und $l_m, r_m \notin H_i$ falls $m \geq 1$.

Wir bezeichnen $l_1 \ldots l_m$ als die vordere Hälfte, $r_m \ldots r_1$ als die hintere Hälfte und k als den Kern von x .

Wir führen nun eine Ordnung auf G ein. Dazu ordnen wir erst die Menge $\{1, \ldots, n\}$ (und damit die Faktoren $H_1, \ldots, H_n$) und in jedem Faktor H_i die Elemente vollständig (wobei 1 das erste Element sei). Dabei werden l und l^{-1} nicht unterschieden. Nun ordnen wir für jedes m die Produkte $l_1 \ldots l_m$ (wobei $l_j \in H_i$ für ein i und die l_j abwechselnd aus verschiedenen Faktoren) lexikographisch. Diese Ordnung in G werde mit $<$ bezeichnet. Nun erweitern wir diese Ordnung $<$ auf die Menge der Paare $\{x, x^{-1}\}$, wobei die Bezeichnung so sei, dass die vordere Hälfte von x bezüglich $<$ vor der von x^{-1} steht.

Dann gelte $\{x, x^{-1}\} \prec \{x', x'^{-1}\}$, wenn entweder $L(x) < L(x')$ oder bei $L(x) = L(x')$ die vordere Hälfte von x echt vor der von x' steht oder falls bei $L(x) = L(x')$ diese übereinstimmen, die vordere Hälfte von x^{-1} vor der von x'^{-1} steht.

Ist $\{x, x^{-1}\} \prec \{x', x'^{-1}\}$, so sagen wir, dass x *vor* x' *steht*.

Ein System $\{x_j\}_{j \in J}$ heisst *kürzer* als ein System $\{x'_j\}_{j \in J}$, wenn $\left\{x_j, x_j^{-1}\right\} \prec \left\{x'_j, x'^{-1}_j\right\}$ für alle $j \in J$, aber für mindestens ein j nicht $\left\{x'_j, x'^{-1}_j\right\} \prec \left\{x_j, x_j^{-1}\right\}$ gilt.

Wir nennen ein System $\{x_j\}_{j \in J}$ *Nielsen-reduziert* (bezüglich $\prec$), wenn es kein zu $\{x_j\}_{j\in J}$ frei äquivalentes System gibt, welches kürzer als $\{x_j\}_{j\in J}$ ist.

HILFSSATZ (1.1) (vgl. [15]). *Ist* $\{x_1, \ldots, x_m\} \subset G$ *ein endliches System, so erhalten wir nach endlich vielen Schritten ein zu* $\{x_1, \ldots, x_m\}$ *frei äquivalentes, Nielsen-reduziertes System* $\{x_1', \ldots, x_m'\}$. □

Sei nun $X = \{x_j\}_{j\in J} \subset G$ eine nicht-leere Teilmenge mit der Eigenschaft, dass jedes $x_j \in X$ endliche Ordnung hat. Wir ersetzen $X = \{x_j\}_{j\in J}$ durch eine (nicht-leere) Menge $X' = \{x_j'\}_{j\in J'} \subset G$ mittels einer (endlichen) Hintereinanderausführung folgender elementarer Transformationen:

(1) Ersetze ein $x_j \in X$ durch $x_j' = x_k^{\varepsilon} x_j x_k^{-\varepsilon}$, $k \neq j$, $\varepsilon = \pm 1$, und lasse die x_i, $i \neq j$, unverändert.

(2) Ersetze ein $x_j \in X$ durch $x_j' = x_j^{-1}$, und lasse die x_i, $i \neq j$, unverändert.

(3) Permutiere in X.

(4) Hat $x_j \in X$ die Ordnung α, $2 \leq \alpha < \infty$, so ersetze x_j durch $x_j' = x_j^{\gamma}$ mit $1 \leq \gamma < \alpha$ und $(\gamma, \alpha) = 1$, und lasse die x_i, $i \neq j$, unverändert.

(5) Erzeugen $x_j, x_k \in X$, $j \neq k$, $1 \neq x_j$, $1 \neq x_k$, eine zyklische Gruppe $\langle x_j' \rangle$, so ersetze x_j durch x_j' und x_k durch $x_k' = 1$, und lasse die x_i, $j \neq i \neq k$, unverändert.

(6) Ist $x_j \in X$, $x_j = 1$, so streiche x_j, und lasse die x_i, $i \neq j$, unverändert.

BEMERKUNG. Es hat auch jedes Element $x_j' \in X'$ endliche Ordnung.

DEFINITION (1.2). (a) Eine nicht-leere Teilmenge $X = \{x_j\}_{j\in J} \subset G$ mit der Eigenschaft, dass jedes $x_j \in X$ endliche Ordnung hat, nennen wir kurz eine E-Menge.

(b) Eine (endliche) Folge von solchen elementaren Transformationen (1) bis (6) nennen wir E-Transformation.

(c) Wir nennen eine E-Menge X' *herleitbar* aus der E-Menge X, wenn es eine E-Transformation von X auf X' gibt.

(d) Eine E-Menge $X = \{x_j\}_{j\in J}$ heisst E-*reduziert*, wenn $x_j \neq 1$ für alle $j \in J$, und wenn es keine aus X herleitbare E-Menge $X' = \{x_j'\}_{j\in J'}$ gibt, für die eine der beiden folgenden Eigenschaften erfüllt ist:

(i) $x'_j = 1$ für ein $j \in J'$;

(ii) $J = J'$, $x'_j \neq 1$ für alle $j \in J'$, und es ist X' kürzer als X .

Als unmittelbare Konsequenz erhalten wir:

HILFSSATZ (1.3). *(a) Die E-Menge X' sei aus der E-Menge X herleitbar. Dann ist $\langle X \rangle = \langle X' \rangle$.*

(b) Sei X eine endliche E-Menge. Dann gibt es eine E-reduzierte, endliche E-Menge X' , die aus X in endlich vielen Schritten herleitbar ist. □

Wir schreiben im folgenden $u_1 \dots u_q \equiv v_1 \dots v_p$ für die Gleichheit zusammen mit der Tatsache, dass die rechte Seite reduziert ist in dem Sinn, dass $L(v_1 \dots v_p) = L(v_1) + \dots + L(v_p)$ gilt (die $u_i, v_j \in G$) .

B. Ist F eine freie Gruppe mit dem freien Erzeugendensystem $B = \{b_i\}_{i \in I}$, so betrachten wir die Elemente aus F als frei reduzierte Worte $w = w(b_1, b_2, \dots)$ relativ zu B .

Ist allerdings $H \subset F$ eine endlich erzeugte Untergruppe von F und $\{x_1, \dots, x_m\}$ ein freies Erzeugendensystem von H , so bedeute die Schreibweise $w_1 = w_1(x_1, \dots, x_m)$ für $w_1 \in H$, dass wir gerade w_1 als frei reduziertes Wort relativ zu $\{x_1, \dots, x_m\}$ betrachten.

C. Wir identifizieren die PSL(2, ℝ) mit der Gruppe aller Automorphismen der oberen Halbebene $\underline{\underline{H}}$.

Es ist PSL(2, ℝ) = SL(2, ℝ)/$\{E, -E\}$, d.h. die PSL(2, ℝ) besteht aus allen Paaren $\{W, -W\}$, $W \in$ SL(2, ℝ) .

Es ruft keine Missverständnisse hervor, wenn wir kurz W statt $\{W, -W\}$ schreiben.

Mit Sp W bezeichnen wir die Spur von $W \in$ SL(2, ℝ) ; in dem oben vereinbarten Sinne können wir (nach geeigneter Festlegung) auch von der Spur eines Elementes der PSL(2, ℝ) sprechen.

Ein Element $W \in$ PSL(2, ℝ) heisst:

(a) *elliptisch*, wenn $|\mathrm{Sp}\, W| < 2$;

(b) *parabolisch*, wenn $W \neq \pm E$, $|\mathrm{Sp}\, W| = 2$ und

(c) *hyperbolisch*, wenn $|\mathrm{Sp}\, W| > 2$.

Ein parabolisches Element hat genau einen Fixpunkt, diesen nennen wir eine *parabolische Spitze*.

Die folgenden Ausführungen sind Standard (vgl. hierzu etwa [3], [4], [6], [7], [9] und [19]).

Unter einer *Fuchsschen Gruppe* verstehen wir hier stets eine endliche erzeugte, diskrete Untergruppe der $PSL(2, \mathbb{R})$, die nicht zyklisch und auch nicht isomorph zur unendlichen Diedergruppe $Z_2 * Z_2$ ist.

Jede Fuchssche Gruppe Γ hat eine Präsentierung der Form:

Erzeugende: $s_1, \ldots, s_m$ (elliptische Elemente);

$p_1, \ldots, p_t$ (parabolische Elemente);

$a_1, b_1, \ldots, a_g, b_g$ (hyperbolische Elemente);

$h_1, \ldots, h_s$ (hyperbolische Randelemente);

Relationen: $s_i^{\alpha_i} = 1$ $(i = 1, \ldots, m)$;

$$\prod_{i=1}^{m} s_i \prod_{j=1}^{t} p_j \prod_{l=1}^{s} h_l \prod_{k=1}^{g} [a_k, b_k] = 1 ,$$

wobei $0 \leq m, t, g, s$, $2 \leq \alpha_i$ $(i = 1, \ldots, m)$ und

$$\mu(\Gamma) .= 2g - 2 + \sum_{i=1}^{m} \left(1 - \frac{1}{\alpha_i}\right) + t + s > 0 .$$

Wir nennen solch ein Erzeugendensystem

$$\{s_1, \ldots, s_m, p_1, \ldots, p_t, h_1, \ldots, h_s, a_1, b_1, \ldots, a_g, b_g\}$$

von Γ , welches die obigen Relationen erfüllt, ein *Standard-Erzeugendensystem von* Γ .

Ist für eine Fuchssche Gruppe Γ der nicht-euklidische Flächeninhalt eines Fundamentalbereiches von Γ endlich (er ist dann gerade $2\pi\mu(\Gamma)$), so nennen wir Γ eine *Fuchssche Gruppe erster Art*.

Eine Fuchssche Gruppe Γ ist genau dann von erster Art, wenn sie ein Standard-Erzeugendensystem besitzt, welches keine hyperbolischen Randelemente enthält, d.h. für welches $s = 0$ ist.

Sei nun Γ eine Fuchssche Gruppe erster Art. Es gelten die Aussagen (die Beweise werden in der Literatur teilweise nur für $s = t = 0$ durchgeführt, eine Übertragung auf den Fall $s = 0$, $t > 0$ ist aber leicht):

(i) Ist $\Gamma_1 \subset \Gamma$ eine Untergruppe von endlichem Index $[\Gamma : \Gamma_1]$, so ist auch Γ_1 eine Fuchssche Gruppe erster Art, und es gilt die Riemann-Hurwitz-Relation

$$[\Gamma : \Gamma_1]\mu(\Gamma) = \mu(\Gamma_1) \ .$$

(ii) Ist $\Gamma_1 \subset \Gamma$ eine Fuchssche Gruppe erster Art, so ist Γ_1 eine Untergruppe von endlichem Index $[\Gamma : \Gamma_1]$ in Γ .

Wir nennen eine Fuchssche Gruppe Γ erster Art *zykloid*, wenn sie genau eine Äquivalenzklasse parabolischer Spitzen (bzgl. Γ) besitzt, d.h. wenn gilt: sind p und q zwei parabolische Erzeugende aus Γ , so gibt es ein $g \in \Gamma$ mit $g\langle p\rangle g^{-1} = \langle q\rangle$.

Eine Fuchssche Gruppe erster Art ist genau dann zykloid, wenn sie ein Standard-Erzeugendensystem besitzt, welches keine hyperbolischen Randelemente und nur ein parabolisches Element enthält, d.h. für welches $s = 0$ und $t = 1$ ist.

2. Gleichungen in freien Produkten

SATZ (2.1). *Sei* $G = H_1 * \dots * H_n$ $(n \geq 2)$ *das freie Produkt der Gruppen* $H_1, \dots, H_n$. *Sei* $1 \neq a_j \in H_j$ *und* p *die Anzahl die* a_j , *dei echte Potenz in* H_j *sind* $(1 \leq j \leq n)$.

Sei $\{x_1, \dots, x_m\} \subset G$ $(m \geq 1)$ *und* H *die von* $x_1, \dots, x_m$ *erzeugte Untergruppe von* G .

Ist $a_1 \dots a_n \quad H$, *so tritt einer der folgenden Fälle ein:*

(2.2) *Es gibt einen freien Übergang von* $\{x_1, \dots, x_m\}$ *zu einem System* $\{y_1, \dots, y_m\}$ *mit* $y_1 = a_1 \dots a_n$.

(2.3) *Es ist* $m \geq 2n-p$, *und es gibt einen freien Übergang von* $\{x_1, \dots, x_m\}$ *zu einem System* $\{y_1, \dots, y_m\}$ *mit* $y_i \in H_j$ $(1 \leq j \leq n \, , \; 1 \leq i \leq 2n-p)$.

Beweis. Wir können annehmen, dass $\{x_1, \dots, x_m\}$ Nielsen-reduziert ist (vgl. Hilfssatz (1.1)). $\{x_1, \dots, x_m\}$ genügt einer Gleichung

$$\prod_{k=1}^{q} x_{\nu_k}^{\varepsilon_k} = a_1 \dots a_n \, , \quad \varepsilon_k = \pm 1 \, , \quad \varepsilon_k = \varepsilon_{k+1} \ \text{falls} \ \nu_k = \nu_{k+1} \, . \tag{2.4}$$

Wir dürfen im folgenden voraussetzen:

(a) $x_i \neq 1$ für alle $i \in \{1, \dots, m\}$;

(b) jedes x_i kommt in (2.4) vor, d.h. $i = \nu_k$ für ein k $(i = 1, \dots, m)$;

(c) q ist minimal unter den Zahlen gewählt, für die eine Gleichung (2.4) gilt.

Nach [15, pp. 4-7] gilt damit stets

$$L\begin{pmatrix} x_{\nu_k}^{\varepsilon_k} \cdots x_{\nu_h}^{\varepsilon_h} \end{pmatrix} \geq L\begin{pmatrix} x_{\nu_l}^{\varepsilon_l} \end{pmatrix} \quad \text{für} \quad 1 \leq k \leq l \leq h \leq q$$

sowie

$$L\begin{pmatrix} x_{\nu_k}^{\varepsilon_k} x_{\nu_{k+1}}^{\varepsilon_{k+1}} x_{\nu_{k+2}}^{\varepsilon_{k+2}} \end{pmatrix} \geq L(x_{\nu_k}) - L(x_{\nu_{k+1}}) + L(x_{\nu_{k+2}}) \quad \text{für} \quad 1 \leq k \leq q-2 .$$

Wir unterscheiden zwei Fälle.

(2.5) Es gibt ein $\lambda \in \{1, \ldots, m\}$, so dass stets $L\begin{pmatrix} x_\nu^\varepsilon x_\lambda x_\mu^\eta \end{pmatrix} > L(x_\nu) - L(x_\lambda) + L(x_\mu)$ ist für $\nu, \mu \in \{1, \ldots, m\}$; $\varepsilon, \eta = \pm 1$ und $\nu \neq \lambda \neq \mu$ oder für $\nu = \lambda$, $\varepsilon = 1$ bzw. $\lambda = \mu$, $\eta = 1$.

Sei etwa $\lambda = \nu_k$ $(1 \leq k \leq q)$. Wir setzen $x_{\nu_{k-1}}^{\varepsilon_{k-1}} =. u_1$, $x_{\nu_k}^{\varepsilon_k} =. u_2$ und $x_{\nu_{k+1}}^{\varepsilon_{k+1}} =. u_3$.

Sei $u_j = l_{j1} \ldots l_{jm_j} k_j r_{jm_j} \ldots r_{j1}$ die symmetrische Normalform von u_j $(j = 1, 2, 3)$. Die $l_{j1}, \ldots, l_{jm_j}, k_j, r_{jm_j}, \ldots, r_{j1}$ nennen wir kurz Stellen von u_j; für eine Stelle von u_j schreiben wir abkürzend einfach v_j $(1 \leq j \leq 3)$.

Dann gibt es ein a_t $(1 \leq t \leq n)$ derart, dass $a_t = v_2$ für eine Stelle v_2 von u_2 oder $a_t = v_1 v_2$ für Stellen v_1 von u_1, v_2 von u_2 oder $a_t = v_2 v_3$ für Stellen v_2 von u_2, v_3 von u_3.

Denn: Angenommen, es gibt ein a_t $(1 \leq t \leq n)$, zu dem alle drei Elemente u_1, u_2, u_3 beitragen. Dann muss u_1 von u_2 die volle vordere, u_3 von u_2 die volle hintere Hälfte wegkürzen, und der Kern k_2 von u_2 muss an der Bildung von a_t beteiligt sein. Dann gilt aber $L(u_1 u_2 u_3) \leq L(u_1) - L(u_2) + L(u_3)$, was einen Widerspruch ergibt.

Als Folgerung daraus erhalten wir: $x_\lambda = x_{\nu_k}$ kommt in (2.4) nur genau einmal vor.

Denn: Angenommen, $x_\lambda = x_{\nu_k}$ kommt in (2.4) zweimal vor, d.h. $\nu_k = \nu_h$ für ein h $(1 \leq h \leq q)$ mit $k \neq h$.

Sei ohne Einschränkung $k < h$.

Da die a_t $(t = 1, \ldots, n)$ alle verschieden sind, ist nach oben notwendig $h = k + 1$, $\varepsilon_h = \varepsilon_k$ und die vordere Hälfte von x_{ν_k} invers zur hinteren Hälfte von x_{ν_k}, d.h. x_{ν_k} ist zu einem Element aus einem Faktor H_s $(1 \leq s \leq n)$ konjugiert.

Dann gilt aber

$$L\left(x_{\nu_k}^3\right) \leq L\left(x_{\nu_k}\right) = L\left(x_{\nu_k}\right) - L\left(x_{\nu_k}\right) + L\left(x_{\nu_k}\right) ,$$

was einen Widerspruch ergibt.

Also kommt $x_\lambda = x_{\nu_k}$ in (2.4) nur genau einmal vor, und es tritt der Fall (2.2) ein.

(2.6) Für jedes $\lambda \in \{1, \ldots, m\}$ gibt es $\nu, \mu \in \{1, \ldots, m\}$, so dass $L\left(x_\nu^\varepsilon x_\lambda x_\mu^\eta\right) \leq L(x_\nu) - L(x_\lambda) + L(x_\mu)$ ist für $\varepsilon, \eta = \pm 1$ und $\nu \neq \lambda \neq \mu$ oder für $\nu = \lambda$, $\varepsilon = 1$ bzw. $\lambda = \mu$, $\eta = 1$.

Nach Hilfssatz (1.7) von [15] ist dann jedes x_λ zu einem Element aus einem H_s konjugiert, und es tritt dann natürlich Fall (2.2) oder Fall (2.3) ein. □

KOROLLAR (2.7) (vgl. auch [2] und [14]). *Sei F eine freie Gruppe mit einem freien Erzeugendensystem B. Seien $B_1, \ldots, B_n$ $(n \geq 2)$ paarweise disjunkte Teilmengen von B; sei F_j die von B_j frei erzeugte Untergruppe von F $(1 \leq j \leq n)$.*

Sei $1 \neq a_j \in F_j$ $(1 \leq j \leq n)$ und $a := a_1 \ldots a_n$.

Sei $H \subset F$ eine endlich erzeugte Untergruppe und $\{x_1, \ldots, x_m\}$ ein freies Erzeugendensystem von H.

Sei $1 \neq w = w(x_1, \ldots, x_m)$ ein (frei reduziertes) Wort in H und α_i die Exponentensumme von x_i in w $(i = 1, \ldots, m)$; es gelte $(\alpha_1, \ldots, \alpha_m) \neq 1$.

Ist $w = a$ in F, so gilt $m \geq n$ und sogar $m \geq 2n$, falls w Element der Kommutatorgruppe H' von H ist.

BEWEIS. Nach Voraussetzung ist $F = F_1 * \ldots * F_n$ das freie Produkt von $F_1, \ldots, F_n$. Wegen $(\alpha_1, \ldots, \alpha_m) \neq 1$ kann w kein primitives Element von H sein. Nun folgt die Behauptung unmittelbar aus Satz (2.1) (dabei ist lediglich zu beachten: Ist K freie Gruppe vom Rang q mit dem freien Erzeugendensystem $\{z_1, \ldots, z_q\}$ und w_1 ein frei reduziertes Wort in K, welches in der Kommutatorgruppe K' von K liegt, so sind mindestens zwei der z_i nötig, um w_1 als Wort in

$z_1, \ldots, z_q$ darzustellen). □

KOROLLAR (2.8) (vgl. auch [14]). *Sei* $G = H_1 * \ldots * H_n$ $(n \geq 2)$ *das freie Produkt der Gruppen* $H_1, \ldots, H_n$. *Sei* $1 \neq a_j \in H_j$ $(1 \leq j \leq n)$. *Sei* $H \subset G$ *eine endlich erzeugte Untergruppe und* $\{x_1, \ldots, x_m\}$ $(m \geq 1)$ *ein minimales Erzeugendensystem von* H, *d.h.* H *kann durch* m, *aber nicht durch* $m-1$ *Elemente erzeugt werden. Sei* $1 \neq w$ *ein Element aus* H, *für welches es eine Darstellung* $w = w(x_1, \ldots, x_m)$ *in* $x_1, \ldots, x_m$ *gibt derart, dass* $(\alpha_1, \ldots, \alpha_m) \neq 1$, *wobei* α_k *die Exponentensumme von* x_k *in der Darstellung* $w = w(x_1, \ldots, x_m)$ *ist* $(k = 1, \ldots, m)$. *Ist* $w = a_1 \ldots a_n$ *in* G, *so gilt* $m \geq n$.

BEWEIS. Angenommen, es gibt keinen freien Übergang von $\{x_1, \ldots, x_m\}$ zu einem System $\{y_1, \ldots, y_m\}$ mit $y_1 \in gH_ig^{-1}$ für ein i $(1 \leq i \leq n)$ und ein $g \in G$. Dann ist H freie Gruppe nach dem Satz von Kurosh mit dem freien Erzeugendensystem $\{x_1, \ldots, x_m\}$. Nach Satz (2.1) gibt es einen freien Übergang von $\{x_1, \ldots, x_m\}$ zu einem System $\{z_1, \ldots, z_m\}$ mit $z_1 = a_1 \ldots a_n = w$, d.h. also $w = w(x_1, \ldots, x_m)$ ist ein primitives Element der freien Gruppe H. Das kann aber wegen $(\alpha_1, \ldots, \alpha_m) \neq 1$ nicht sein, und unsere Annahme ist falsch. Sei also ohne Einschränkung etwa $x_1 \in H_1$. Sei N der von H_1 in G erzeugte Normalteiler und $\phi: G \to G/N \cong H_2 * \ldots * H_n$ der kanonische Epimorphismus.

Es ist $\phi(x_1) = \phi(a_1) = 1$ und $\phi(w) = \phi(a_1 \ldots a_n) = \phi(a_2) \ldots \phi(a_n)$.

Nun folgt induktiv $m-1 \geq n-1$, d.h. $m \geq n$. □

BEMERKUNG. Der Zusatz in Korollar (2.7) für den Fall, dass w in der Kommutatorgruppe H' von H liegt, lässt sich nicht so ohne weiteres für allgemeinere freie Produkte beweisen.

Als direkte Folgerung aus dem Beweis von Satz (2.1) erhalten wir noch

KOROLLAR (2.9) *Sei* $G = \langle s_1, \ldots, s_n \mid s_1^{\alpha_1} = \ldots = s_n^{\alpha_n} = 1 \rangle \cong Z_{\alpha_1} * \ldots * Z_{\alpha_n}$, $2 \leq n$, $2 \leq \alpha_i$ *für* $i = 1, \ldots, n$.

Sei $\{x_1, \ldots, x_m\} \subset G$ $(m \geq 1)$ *und* H *die von* $x_1, \ldots, x_m$ *erzeugte Untergruppe von* G. *Ist* $s_1 \ldots s_n \in H$, *so tritt einer der folgenden Fälle ein:*

(2.10) *Es gibt einen freien Übergang von* $\{x_1, \ldots, x_m\}$ *zu einem System* $\{y_1, \ldots, y_m\}$ *mit* $y_1 = s_1 \ldots s_n$.

(2.11) *Es ist* $H = G$. □

Im Falle einer freien Gruppe F können wir Satz (2.1) erweitern, indem wir $a_1 \ldots a_n$ durch eine Potenz $(a_1 \ldots a_n)^\alpha$, $\alpha \neq 0$, ersetzen.

SATZ (2.12) *Sei* F *eine freie Gruppe mit einem freien Erzeugendensystem* B. *Seien* $B_1, \ldots, B_n$ $(n \geq 2)$ *paarweise disjunkte Teilmengen von* B; *sei* F_j *die von* B_j *frei erzeugte Untergruppe von* F $(1 \leq j \leq n)$.

Sei $1 \neq a_j \in F_j$ *und* p *die Anzahl der* a_j, *die echte Potenz in* F_j *sind* $(1 \leq j \leq n)$. *Sei* $\{x_1, \ldots, x_m\} \subset G$ $(m \geq 1)$ *und* H *die von* $x_1, \ldots, x_m$ *erzeugte Untergruppe von* F.

Ist $(a_1 \ldots a_n)^\alpha \in H$ *für ein* $\alpha \neq 0$, *so tritt einer der folgenden Fälle ein:*

(2.13) *Est gibt einen freien Übergang von* $\{x_1, \ldots, x_m\}$ *zu einem System* $\{y_1, \ldots, y_m\}$ *mit* $y_1 = (a_1 \ldots a_n)^\beta$, $\beta \geq 1$ *und* β *ist die kleinste positive Zahl, für die eine Beziehung* $(a_1 \ldots a_n)^\beta \in H$ *gilt.*

(2.14) *Es ist* $m \geq 2n-p$, *und es gibt einen freien Übergang von* $\{x_1, \ldots, x_m\}$ *zu einem System* $\{y_1, \ldots, y_m\}$ *mit* $y_i \in H_j$ $(1 \leq j \leq n, 1 \leq i \leq 2n-p)$.

BEWEIS. Wir können annehmen, dass $\{x_1, \ldots, x_m\}$ ein freies Erzeugendensystem von H ist. Es trete nun nicht der Fall (2.13) ein, und es sei schon α die kleinste positive Zahl, für die eine Beziehung $(a_1 \ldots a_n)^\alpha \in H$ gilt.

Es ist $\alpha = 1$.

Denn: Angenommen $\alpha \geq 2$. $\{x_1, \ldots, x_m\}$ genügt einer Gleichung $w(x_1, \ldots, x_m) = (a_1 \ldots a_m)^\alpha$. Da der Fall (2.13) nicht eintritt, ist $w(x_1, \ldots, x_m)$ kein primitives Element von H. Nach [1] ist damit $w(x_1, \ldots, x_m)$ eine echte Potenz in H. Das ist ein Widerspruch zur Minimalität von α.

Also ist $\alpha = 1$. Nun folgt (2.12) aus Satz (2.1). □

BEMERKUNG. Satz (2.12) lässt sich nicht erweitern für beliebige freie Produkte $G = H_1 * \ldots * H_n$ von Gruppen $H_1, \ldots, H_n$, nicht einmal für freie Produkte $G = Z_{\alpha_1} * \ldots * Z_{\alpha_n}$, $2 \leq n$, die $\alpha_i \geq 2$, endlicher zyklischer Gruppen (vgl. hierzu etwa Satz 1 aus [12]). Für Untergruppen freier Produkte zyklischer Gruppen, die von Elementen endlicher Ordnung erzeugt werden, ist eine gewisse Erweiterung von Satz (2.12) aber möglich (vgl. dazu auch [5], [12], und [18]). Hier ergänzen wir die in [5] und [18] gemachten Untersuchungen durch den folgenden Satz.

SATZ (2.15) *Sei* $G = \langle s_1, \ldots, s_n | s_1^{\alpha_1} = \ldots s_n^{\alpha_n} = 1 \rangle \cong Z_{\alpha_1} * \ldots * Z_{\alpha_n}$, $2 \le n$, *die* α_i *Primzahlen* $(i = 1, \ldots, n)$; *und es sei* $\alpha_1 + \alpha_2 \ge 5$ *falls* $n = 2$ *ist, d.h.* G *sei nicht isomorph zur unendlichen Diedergruppe.*

Seien $x_1, \ldots, x_m$ $(m \ge 1)$ *Elemente endlicher Ordnung in* G, *und sei* H *die von* $x_1, \ldots, x_m$ *erzeugte Untergruppe von* G.

Genau dann hat H *in* G *endlichen Index* $[G:H]$, *wenn* $y(s_1 \ldots s_n)^{\alpha} y^{-1} \in H$ *für ein* $\alpha \neq 0$ *und ein* $y \in G$.

Darüberhinaus gilt: Hat H *in* G *den endlichen Index* $[G:H]$ *und ist* α *die kleinste positive Zahl mit* $y(s_1 \ldots s_n)^{\alpha} y^{-1} \in H$ *für ein* $y \in G$, *so ist* $[G:H] = \alpha$, *und ein Repräsentantsystem für die Rechstrestklassen von* G *nach* H *ist gegeben durch* $\{1, s_1 \ldots s_n, \ldots, (s_1 \ldots s_n)^{\alpha-1}\}$.

BEWEIS. Hat H in G endlichen Index $[G:H]$, so ist natürlich $y(s_1 \ldots s_n)^{\alpha} y^{-1} \in H$ für ein $\alpha \neq 0$ und ein $y \in G$.

Wir beweisen nun die Umkehrung.

Es sei also $y(s_1 \ldots s_n)^{\alpha} y^{-1} \in H$ für ein $\alpha \neq 0$ und ein $y \in G$. Unter α verstehen wir im folgenden die kleinste positive Zahl, für die solch eine Beziehung gilt.

Wir dürfen — eventuell nach Ersetzen von x_i durch yx_iy^{-1} — voraussetzen, dass schon $(s_1 \ldots s_n)^{\alpha} \in H$ gilt.

Wir bemerken, dass jedes x_i als Element endlicher Ordnung in G konjugiert ist zu einer Potenz von einem s_j.

In der Bezeichnungsweise von §1 ist $\{x_1, \ldots, x_m\}$ eine E-Menge in G. Wir können annehmen, dass die E-Menge $\{x_1, \ldots, x_m\}$ E-reduziert ist (vgl. Hilfssatz (1.3)).

Dann gilt insbesondere $x_i \neq 1$ für alle i $(1 \le i \le m)$, keine zwei der x_i erzeugen eine zyklische Gruppe, H ist das freie Produkt $H = \langle x_1 \rangle * \ldots * \langle x_m \rangle$ der zyklischen Gruppen $\langle x_i \rangle$ $(i = 1, \ldots, m)$ und $L(x_i^{\varepsilon} x_j^{\eta}) \ge L(x_i)$, $L(x_j)$; $\varepsilon, \eta = \pm 1$, $i \neq j$ $(1 \le i, j \le m)$ (vgl. hierzu etwa [5] und [18]).

Es gilt aber auch $L(x_i^{\varepsilon} x_j x_k^{\eta}) > L(x_i) - L(x_j) + L(x_k)$ für $\varepsilon, \eta = \pm 1$ und $i \neq j \neq k$ $(1 \le i, j, k \le m)$.

Denn: Angenommen, es gibt $i \neq j \neq k$ und $\varepsilon, \eta = 1$ mit

$L(x_i^\varepsilon x_j x_k^\eta) \le L(x_i) - L(x_j) + L(x_k)$.

Da $\{x_1, \ldots, x_m\}$ E-reduziert ist, werden beide Hälften von x_j geschluckt, und es liegt für ein λ $(1 \le \lambda \le n)$ folgende Situation vor:

$$x_j \equiv v s_\lambda^\lambda v^{-1} , \quad 1 \le \gamma < \alpha_\lambda , \quad x_i^\varepsilon \equiv u_i s_\lambda^{\delta_i} v^{-1} , \quad 1 \le \delta_i < \alpha_\lambda ,$$

$$x_k^\eta \equiv v s_\lambda^{\delta_k} u_k , \quad 1 \le \delta_k < \alpha_\lambda .$$

Da α_λ eine Primzahl ist, gibt es ein δ mit $\delta\lambda \equiv 1 \pmod{\alpha_\lambda}$, und es ist $x_i^\varepsilon x_j^{-\delta_i\delta} = u_i v^{-1}$ und $L(x_i^\varepsilon x_j^{-\delta_i\delta}) < L(x_i)$ im Widerspruch dazu, dass $\{x_1, \ldots, x_m\}$ E-reduziert ist, denn aus $L(x_i^\varepsilon x_j^{-\delta_i\delta}) < L(x_i)$ folgt $L(x_j^{\delta_i\delta} x_i^\varepsilon x_j^{-\delta_i\delta}) < \max\{L(x_i), L(x_j)\}$ wegen $i \ne j$ (vgl. etwa [5] und [18]).

(2.16) $L(x_i^\varepsilon x_j x_k^\eta) > L(x_i) - L(x_j) + L(x_k)$ für $\varepsilon, \eta = \pm 1$ und $i \ne j \ne k$ $(1 \le i, j, k \le m)$.

$\{x_1, \ldots, x_m\}$ genügt einer Gleichung

(2.17) $\prod_{k=1}^{q} x_{\nu_k}^{\varepsilon_k} = (s_1 \ldots s_n)^\alpha$ mit $\varepsilon_k = \pm 1$, $\varepsilon_k = \varepsilon_{k+1}$ falls $\nu_k = \nu_{k+1}$, wobei wir wider voraussetzen, dass q minimal unter den Zahlen gewählt ist, für die eine Gleichung (2.17) mit dem minimalen α gilt.

Insbesondere haben wir in (2.17) keine Situation: $x_{\nu_k} = \ldots = x_{\nu_{k+l}}$ und $x_{\nu_k}^{l+1} = 1$ $(1 \le k \le k+l \le q)$.

Es gebe in (2.17) zwei verschiedene Indizes j, i $(1 \le j < i \le q)$ mit $\nu_j = \nu_i = k$, $\varepsilon_j = \varepsilon_i$ und $\nu_h \ne k = \nu_j$ für $j < h < i$.

Wir wollen zeigen, dass $j = i-1$ ist .

Es ist möglicherweise $\nu_{j-1} = k$ und (oder) $\nu_{i+1} = k$.

Da die α_λ aber Primzahlen sind, können wir (indem wir eventuell die Ordnung in den Faktoren Z_{α_λ} anders festlegen und x_k durch eine geeignete Potenz ersetzen) ohne Einschränkung annehmen, dass $\nu_{i+1} \ne k$ ist.

Sei also nun $\nu_j = \nu_i = k$, $\varepsilon_j = \varepsilon_i$, $\nu_h \ne k$ für $j < h < i$, $\nu_{i+1} \ne k$, aber eventuell $\nu_{j-1} = k$.

Sei ohne Einschränkung $\varepsilon_j = \varepsilon_i = 1$.

Es ist $x_k \equiv u s_\lambda^\beta u^{-1}$ $(1 \le \lambda \le n)$ und

$$x_k \left(\prod_{p=j+1}^{i-1} x_{\nu_p}^{\varepsilon_p}\right) x_k = us_\nu^\beta u^{-1} \left(\prod_{p=j+1}^{i-1} x_{\nu_p}^{\varepsilon_p}\right) us_\lambda^\beta u^{-1}$$

$$= us_\lambda^{\beta'} (s_\lambda \dots s_n s_1 \dots s_{\lambda-1})^\delta s^{\beta''} u^{-1} , \quad \delta \geq 0 ,$$

da $\{x_1, \dots, x_m\}$ E-reduziert ist.

Wir betrachten diese obige Geichung.

Angenommen $\delta > 0$. Mit Hilfe von (2.16) erhalten wir dann:

Wegen $\nu_{i+1} \neq k$ ist notwendig $\beta'' \equiv \beta \equiv 1 \pmod{\alpha_\lambda}$;

wegen $\nu_h \neq k$ für $j < h < i$ ist damit weiter $\beta' + 1 \equiv \beta \equiv 1 \pmod{\alpha_\lambda}$, d.h.

$\beta' \equiv 0 \pmod{\alpha_\lambda}$.

Daher erhalten wir

$$s_\lambda u^{-1} \left(\prod_{p=j+1}^{i-1} x_{\nu_p}^{\varepsilon_p}\right) us_\lambda = s_\lambda s_{\lambda+1} \dots s_n (s_1 \dots s_n)^{\delta-1} s_1 \dots s_{\lambda-1} s_\lambda$$

also

$$\left(\prod_{p=j+1}^{i-1} x_{\nu_p}^{\varepsilon_p}\right) x_k = us_\lambda^{-1} \dots s_1^{-1} (s_1 \dots s_n)^\sigma s_1 \dots s_\lambda u_\lambda^{-1} .$$

Eine solche Gleichung widerspricht aber der Minimalität von α falls $0 < \delta < \alpha$ ist, sonst aber bei $\delta = \alpha$ der von q .

Also ist $\delta = 0$ und damit $j = i-1$ wegen (2.16).

Da die x_i alle endliche Ordnung haben, erhalten wir daher – eventuell nach geigneter Umnummerierung und da wir annehmen können, dass nun jedes x_i in (2.17) vorkommt – aus Gleichung (2.17) eine Gleichung der Form

(2.18) $x_1^{\gamma_1} \dots x_m^{\gamma_m} = (s_1 \dots s_n)^\alpha$ mit $x_i \neq x_j$ für $i \neq j$ und $x_i^{\gamma_i} \neq 1$ $(1 \leq i, j \leq m)$.

Es ist $x_i \equiv u_i s_j^{\beta_i} u_i^{-1}$ $(1 \leq i \leq m)$ für ein j $(1 \leq j \leq n)$.

Da die α_j Primzahlen sind, gibt es zu x_i eine Zahl ρ_i und ein $y_i \in G$ mit $\rho_i \gamma_i \equiv 1 \pmod{\alpha_j}$ und $x_i = y_i^{\rho_i}$ (beachte: x_i und y_i erzeugen dieselbe zyklische Gruppe).

Indem wir eventuell x_i durch y_i ersetzen, können wir also für die folgenden Überlegungen ohne Einschrankung $\gamma_i = 1$ für alle i annehmen, d.h.

(2.19) $x_1 \dots x_m = (s_1 \dots s_n)^\alpha$ mit $x_i \neq x_k$ für $i \neq k$ und $x_i \equiv u_i s_j^{\beta_i} u_i^{-1}$ $(1 \leq i \leq m)$ für ein j $(1 \leq j \leq n)$.

Mittels (2.16) erhalten wir nun notwendig

(2.20) $x_i \equiv u_i s_j u_i^{-1}$ $(1 \le i \le m)$, d.h. $\beta_i \equiv 1 \pmod{\alpha_j}$, für ein j $(1 \le j \le n)$.

Sei nun für den Rest des Beweises G treu dargestellt als zykloide Fuchssche Gruppe erster Art derart, dass

(1) $s_1, \ldots, s_n$ elliptische Elemente sind;

(2) $p .= (s_1 \ldots s_n)^{-1}$ parabolisches Element is und

(3) $\{s_1, \ldots, s_n, p\}$ eine Standard-Erzeugendensystem von G ist (vgl. etwa [7, pp.247-249]; hier wird $\alpha_1 + \alpha_2 \ge 5$ falls $n = 2$ benötigt).

Als Untergruppe von G ist H auch diskret; natürlich ist H nicht zyklisch, wegen $\alpha_1 + \alpha_2 \ge 5$ falls $n = 2$ ist H auch nicht isomorph zur unendlichen Diedergruppe, d.h. H ist Fuchssche Gruppe. Da H das freie Produkt $H = \langle x_1 \rangle * \cdots * \langle x_m \rangle$ der zyklischen Gruppen $\langle x_i \rangle$ ist, hat H den Rang m, d.h. also H kann von m, aber nicht von m-1 Elementen erzeugt werden.

Andererseits gibt es für H ein Standard-Erzeugendensystem. Dieses muss wegen $H = \langle x_1 \rangle * \cdots * \langle x_m \rangle$ mindestens m Elemente endlicher Ordnung (elliptische Elemente) enthalten; da mit $s_1 \ldots s_n$ auch $(s_1 \ldots s_n)^\alpha$ parabolisch ist, muss es auch mindestens ein parabolisches Element enthalten. Es kann aber auch nicht mehr als m+1 Elemente enthalten, da H den Rang m hat und nach dem Satz von Gruschko der Rang von H direkt aus einem Standard-Erzeugendensystem ablesbar ist.

Also ist wegen (2.19) und (2.20) schon $\{x_1, \ldots, x_m, p_1\}$, $p_1 .= p^\alpha = (s_1 \ldots s_m)^{-\alpha}$ ein Standard-Erzeugendensystem von H. Damit ist H eine Fuchssche Gruppe erster Art, hat also in G endlichen Index $[G:H]$.

Wir haben noch die zusätzliche Aussage zu zeigen. Es habe also H in G den endlichen Index $[G:H]$, und es sei α die kleinste positive Zahl mit $y(s_1 \ldots s_n)^\alpha y^{-1} \in H$ für ein $y \in G$. Wir nehmen wieder an, dass schon $(s_1 \ldots s_n)^\alpha \in H$ gilt (es zeigt sich, dass dies keine Einschränkung bedeutet).

Ist $[G:H] = 1$, so ist nichts zu zeigen.

Sei nun $[G:H] > 1$ und $a \in G \backslash H$.

Wir bilden $a(s_1 \ldots s_n)a^{-1}$. Da H in G den endlichen Index $[G:H]$ hat, gibt es eine positive Zahl ϕ_1 mit $a(s_1 \ldots s_n)^{\phi_1} a^{-1} \in H$. Da H nur eine Äquivalenzklasse parabolischer Spitzen (bzgl. H) besitzt (den obigen Ausführungen entnehmen wir, dass H zykloid ist), gibt es ein $b \in H$ und eine Zahl $\phi_2 \neq 0$ mit $ba(s_1 \ldots s_n)^{\phi_1} a^{-1} b^{-1} = (s_1 \ldots s_n)^{\phi_2}$. Wegen $\alpha_1 + \alpha_2 \geq 5$ falls $n = 2$ folgt $\phi_1 = \phi_2$ (vgl. die Hilfssätze 2 und 3 aus [17]; dies folgt aber auch fast unmittelbar aus der Diskretheit von G). Wegen $b \in H$ und $(s_1 \ldots s_n)^\alpha \in H$ gibt es also ein $d \in H$ und eine Zahl ϕ mit $1 \leq \phi \leq \alpha$ derart, dass $a = d(s_1 \ldots s_n)^\phi$, d.h. $Ha = H(s_1 \ldots s_n)^\phi$. □

Damit ist alles gezeigt.

BEMERKUNGEN. (1) Die Voraussetzung in Satz (2.15), dass H endlich erzeugt ist, ist nicht wesentlich. Denn: Sei $G = \langle s_1, \ldots, s_n \mid s_1^{\alpha_1} = \ldots = s_n^{\alpha_n} = 1 \rangle$, $2 \leq n$, die α_i Primzahlen; sei $\alpha_1 + \alpha_2 \geq 5$ falls $n = 2$ ist.

Seien $x_1, x_2, \ldots$ (eventuell unendlich viele) Elemente endlicher Ordnung in G , und sei H die von $x_1, x_2, \ldots$ erzeugte Untergruppe von G . Ist $y(s_1 \ldots s_n)^\alpha y^{-1} \in H$ für ein $\alpha \neq 0$ und ein $y \in G$, so genügen endlich viele der x_i , etwa $x_1, \ldots, x_m$, um $y(s_1 \ldots s_n)^\alpha y^{-1}$ als Element von H darzustellen; und es ist die Untergruppe $\langle x_1, \ldots, x_m \rangle$ dann von endlichem Index in G , also erst recht H , d.h. aber H ist endlich erzeurbar.

(2) Es wäre wichtig zu wissen, ob Satz (2.15) auch für beliebige $\alpha_i \geq 2$ $(i = 1, \ldots, n)$ gilt.

(3) Für die rationale Modulgruppe

$$\Gamma := \left\{ A : z \mapsto \frac{az+b}{cz+d} \,\middle|\, z \in \mathbb{C} \cup \{\infty\} \ ; \quad a, b, c, d \in \mathbb{Z} \ , \quad ad-bc = 1 \right\} = \mathrm{PSL}(2, \mathbb{Z})$$

ergeben sich im Zusammenhang mit Satz (2.15) interessante Fragestellungen.

Γ ist zykloide Fuchssche Gruppe erster Art; Γ wird erzeugt durch die Elemente $T : z \mapsto -\frac{1}{z}$ sowie $R : z \mapsto -\frac{1}{z+1}$, und es ist $U := TR : z \mapsto z+1$. Es gelten die definierenden Relationen $T^2 = T^3 = 1$, und es ist $\Gamma = \langle T \rangle * \langle R \rangle$, d.h. Γ ist das freie Produkt der zyklischen Gruppen $\langle T \rangle$ und $\langle R \rangle$ (vgl. etwa [16]).

Under der *Hauptkongruenzgruppe* $\Gamma[N]$ der Stufe N , N eine natürliche Zahl, verstehen wir die Untergruppe

$$\Gamma[N] := \left\{ A : z \mapsto \frac{az+b}{cz+d} \,\middle|\, A \in \Gamma \ , \ \begin{pmatrix} a & b \\ c & d \end{pmatrix} \equiv \begin{pmatrix} 1 & 0 \\ 0 & 1 \end{pmatrix} \pmod N \right\} \subset \Gamma \ .$$

Eine Untergruppe $\Gamma_1 \subset \Gamma$ heisst *Kongruenzgruppe*, wenn $\Gamma[N] \subset \Gamma_1$ ist für ein $N \geq 1$

(Γ_1 hat dann natürlich endlichen Index $[\Gamma:\Gamma_1]$ in Γ).

Für zykloide Untergruppen von Γ gelten die folgenden Existenzsätze:

(i) Zu jedem System ganzer Zahlen $\mu \geq 1$ und $g, e_2, e_3 \geq 0$, welches die Riemann-Hurwitz-Relation $\mu = 6(2g-1)+3e_2+4e_3$ erfüllt, gibt es eine zykloide Untergruppe Γ_1 von Γ mit

(a) μ ist der Index $[\Gamma:\Gamma_1]$ von Γ_1 in Γ ;

(b) g ist das Geschlecht von Γ_1 , d.h. das Geschlecht des Quotientenraumes $\underline{\underline{H}}^*/\Gamma_1$, wobei $\underline{\underline{H}}$ die obere Halbebene und $\underline{\underline{H}}^* = \underline{\underline{H}} \cup \mathbb{Q} \cup \{\infty\}$ ist;

(c) e_k , $k = 2$ oder 3 , ist die Anzahl der Konjugationsklassen zyklischer Gruppen der Ordnung k in Γ_1 (vgl. hierzu [10] und [11]).

(ii) Es gibt nur endlich viele zykloide Kongruenzgruppen in Γ . Der Index $\mu = [\Gamma:\Gamma_1]$ einer zykloiden Kongruenzgruppe $\Gamma_1 \subset \Gamma$ ist ein Teiler von $\bar{N} := 2^4.3^2.5.7.11 = 55440$. Zu jedem positiven Teiler μ von $\bar{N}$ gibt es zykloide Kongruenzgruppen vom Index μ in Γ . Zu $\mu = 1, 2, 3, 4, 5, 6, 7, 8, 9, 10, 11, 12, 21$ gibt es zykloide Kongruenzgruppen vom Index μ und Geschlecth null in Γ (vgl. hierzu [13]).

Es gibt also nur relativ wenige zykloide Kongruenzgruppen in Γ . Petersson führt aus, dass mit Hilfe des in [13] erläuterten Konstruktionsverfahrens die explizite Bestimmung aller zykloiden Kongruenzgruppen in Γ faktisch ausführbar ist, wenn auch sehr mühsam. Er selbst bestimmt in [13] explizit alle zykloiden Kongruenzgruppen in Γ , die mit den Durschschnitten zykloider Kongruenzgruppen von (nichttrivialem) Primzahlpotenzindex in Γ zusammenfallen.

Seien nun $Z_1, Z_2, \ldots$ Elemente endlicher Ordnung (d.h. der Ordnung 2 oder 3) in Γ , und sei Γ_1 die von $Z_1, Z_2, \ldots$ erzeugte Untergruppe von Γ .

Ferner sei $AU^{\alpha}A^{-1} \in \Gamma_1$ für ein $\alpha \neq 0$ und ein $A \in \Gamma$, wobei wieder $U = TR : z \mapsto z+1$ ist.

Dem Beweis von Satz (2.15) und der anschliessenden Bermerking (1) entnehmen wir: Γ_1 ist zykloide Untergruppe von endlichem Index $[\Gamma:\Gamma_1]$ in Γ und vom Geschlecht null.

Weiter ist Γ_1 kein Normalteiler von Γ falls der Index $[\Gamma:\Gamma_1]$ von Γ_1 in Γ grösser als 6 ist (dies folgt daraus, dass es auf Γ genau 6 gerade abelsche Charaktere gibt).

Mittels Satz (2.15) und des Konstruktionsverfahrens aus [13] lässt sich nun bestimmen, ob unsere Untergruppe Γ_1 Kongruenzgruppe in Γ ist oder nicht.

Literatur

[1] G. Baumslag, "Residual nilpotence and relations in free groups", *J. Algebra* 2 (1965), pp. 271-282, MR31:3487.

[2] C.C. Edmunds, "Representing products of disjoint words in a free group", *Illinois J. Math.* 25 (1981), pp. 589-592, MR82j:20057.

[3] A.H.M. Hoare, A. Karrass, and D. Solitar, "Subgroups of finite index of Fuchsian groups", *Math. Z.* 120 (1971), pp. 289-298, MR44:2837.

[4] A.H.M. Hoare, A. Karrass, and D. Solitar, "Subgroups of infinite index in Fuchsian groups", *Math. Z.* 125 (1972), pp. 59-69, MR45:2029.

[5] R.N. Kalia and G. Rosenberger, "Über Untergruppen ebener diskontinuierlicher Gruppen", *Festschrift zum 65. Geburtstag von R.C. Lyndon* (erscheint).

[6] L. Keen, "Canonical polygons for finitely generated Fuchsian groups", *Acta Math.* 115 (1965), pp. 1-16, MR32:1349.

[7] J. Lehner, *Discontinuous Groups and Automorphic Functions* (A.M.S. Surveys 8, Amer. Math. Soc., 1964), MR29:1332.

[8] R.C. Lyndon and P.E. Schupp, *Combinatorial Group Theory* (Ergibnisse der Math. 189, Springer, Berlin, Heidelberg, New York, 1977), MR58:28182.

[9] A.M. Mcbeath, "Fuchsian groups", *Proc. of the Dundee Summer School* (1961).

[10] M.H. Millington, "On cycloidal subgroups of the modular group", *Proc. London Math. Soc.* (3) 19 (1969), pp. 164-176, MR40:1484.

[11] M.H. Millington, "Subgroups of the classical modular group", *J. London Math. Soc.* (2) 1 (1969), pp. 351-357, MR39:5477.

[12] N. Peczynski, G. Rosenberger, and H. Zieschang, "Über Erzeugende ebener diskontinuierlicher Gruppen", *Inventiones math.* 29 (1975), pp. 161-180, MR52:3340.

[13] H. Petersson, "Über die Konstruktion zykloider Kongruenzgruppen in der rationalen Modulgruppe", *J. reine und angew. Math.* 250 (1971), pp. 182-212, MR45:5355.

[14] G. Rosenberger, "Über alternierende Wörter in freien Produkten mit Amalgam", *Archiv der Math.* 31 (1978), pp. 417-422, MR80h:20047.

[15] G. Rosenberger, "Gleichungen in freien Produkten mit Amalgam", *Math. Z.* 173 (1980), pp. 1-11, MR81k:20038; Berichtigung, *Math. Z.* 178 (1981), p. 579, MR82k:20050.

[16] B. Schoeneberg, *Elliptic Modular Functions* (Grundlehren der math. Wiss. 203, Springer, Berlin, Heidelberg, New York, 1974), MR54:236.

[17] H. Zieschang, "Über die Nielsensche Kürzungsmethode in freien Produkten mit Amalgam", *Inventiones math.* 10 (1970), pp. 4-37, MR41:8528.

[18] H. Zieschang, "On subgroups of free products of cyclic groups" (in Russian), to appear.

[19] H. Zieschang, E. Vogt, and H.-D. Coldewey, *Surfaces and Planar Discontinuous Groups* (Lecture Notes in Math. 835, Springer, Berlin, Heidelberg, New York, 1980), MR82h:57002.

Abteilung Mathematik,
Universität Dortmund,
D-4600 Dortmund 50, Federal Republic of Germany

PROCEEDINGS OF 'GROUPS — KOREA 1983'
KYOUNGJU, August 1983

20D15,
16A22

ON THE CIRCLE GROUPS OF FINITE NILPOTENT RINGS

Ken-Ichi Tahara and Akinori Hosomi

1. Introduction

Let R be a ring, and consider the following operation $\circ$ in R :

$$a \circ b = a + b + a \times b = a + b + ab ,$$

where $+$ and $\times$ denote the addition and multiplication in R . If R makes a group under the operation $\circ$, the group is called the circle group of R and will be denoted by $(R, \circ)$. It is clear that a nilpotent ring R of index $c+1$ has the circle group which is nilpotent and of class at most c . But conversely not every nilpotent group arises necessarily as the circle group of a nilpotent ring. It is of interest to know which nilpotent groups of class c arise as the circle groups of nilpotent rings of index $c+1$. Hales and Passi [3] and Ault and Watters [1] gave criteria for a nilpotent group of class 2 to be the circle group of a nilpotent ring of index 3 .

In this paper we give a criterion for a finite nilpotent group of class 3 to be the circle group of a nilpotent ring of index 4 . On the other hand, Kruse and Price [4] proved any nilpotent group of class 3 and of order p^4 is not the circle group of any nilpotent ring of index 4 . But their method can not be available for p-groups of order p^5. Now as its application of the main theorem, we can show that there exist p-groups of order p^5 and of nilpotent class 3 which are the circle groups of nilpotent rings of index 4 , and moreover we can classify all p-groups of order p^5 and of nilpotent class 3 which are the circle groups of nilpotent rings of index 4 .

2. Preliminaries

We state easy lemmas. The first one is concerned with calculations of commutators of the circle groups:

LEMMA 2.1. *Let G be the circle group of a ring R , and denote by g' the inverse element of an element g of G . Then for any elements g and h of G ,*

1) $g' \circ h \circ g = h + g'h + hg + g'hg$,

2) $[g, h] = g'h' + h'g + g'h'g + g'h'h + h'gh + g'h'gh$.

COROLLARY 2.2. *Let G be the circle group of a nilpotent ring R of index* 4. *Denote by g' the inverse element of an element g of G. Then we have $g' = -g + g^2 - g^3$, and for any elements g and h of G,*

1) $g' \circ h \circ g = h + (hg - gh) + (g^2h - ghg)$,

2) $[g, h] = (gh - hg) - (g^2h - ghg) + (h^2g - hgh)$.

The second one is the following (cf. [5, Lemma 1.1]):

LEMMA 2.3. *Let G be the circle group of a ring R, and S be a two-sided ideal of R. Then $(S, \circ)$ makes a normal subgroup of $G = (R, \circ)$, and $G/(S, \circ)$ is isomorphic to the circle group $(R/S, \circ)$ of R/S.*

Proof. Though its proof is easy and given in [5], we shall prove it for its completeness. For any element s of S, there exists an element s' in R such that $s \circ s' = s + s' + ss' = 0$, and hence $s' = -s - ss' \in S$. For any elements s and t of S, $s \circ t = s + t + st \in S$. Thus $(S, \circ)$ is a subgroup of G. For any element s of $(S, \circ)$ and any element g of G, $g' \circ s \circ g = s + g's + sg + g'sg \in (S, \circ)$ by Lemma 2.1. Thus $(S, \circ)$ is normal in G. Next we claim that $(R/S, \circ)$ is a group. For any element $g + S$ of $(R/S, \circ)$, denote by g' the inverse element of g in G, then $(g+S) \circ (g'+S) = (g'+S) \circ (g+S) = S$. Thus $g'+S$ is the inverse element of $g+S$ in $(R/S, \circ)$. Moreover if we define a map $\phi : G \to (R/S, \circ)$ by $g \mapsto g+S$, it is easy to show that ϕ is a surjective homomorphism and $\mathrm{Ker}\,\phi = (S, \circ)$. Therefore $G/(S, \circ)$ is isomorphic to $(R/S, \circ)$. □

Now let G be the circle group of a ring R, and $\mathbb{Z}G$ be the group ring of G over $\mathbb{Z}$, the ring of all rational integers. Then we denote by $I(G)$ the augmentation ideal of $\mathbb{Z}G$. Now define a map $\rho_G : \mathbb{Z}G \to R$ by $\Sigma a \cdot g \mapsto \Sigma ag$, $\rho_G(1_G) = 0_R$, where 1_G is the identity of G, and 0_R is the zero of R. We denote $\rho_G|I(G)$ by the notation ρ_G. Then we have easily

PROPOSITION 2.4. $\rho_G : I(G) \to R$ *is a surjective homomorphism of rings.* □

Let G be any group. We denote by $G = D_1(G) \supseteq D_2(G) \supseteq \cdots \supseteq D_n(G) \supseteq \cdots$ and $G = G_1 \supseteq G_2 \supseteq \cdots \supseteq G_n \supseteq \cdots$ the series of dimension subgroups of G over $\mathbb{Z}$ and the lower central series of G, respectively.

COROLLARY 2.5. *Let G be the circle group of a ring R. Then it follows that, for any $n \geq 1$,*

$$(R^n, \circ) \supseteq D_n(G) \supseteq G_n .$$

Proof. Let g be any element of $D_n(G)$. Then $g - 1$ is contained in $I^n(G)$, and hence $g = \rho_G(g-1) \in \rho_G(I^n(G)) = R^n$. Therefore $(R^n, \circ) \supseteq D_n(G) \supseteq G_n$. □

Let G be a group, N be a normal subgroup of G and $\phi_N : G \to G/N$ be the canonical homomorphism from G to G/N. We can extend it $\mathbb{Z}$-linearly to $\mathbb{Z}G \to \mathbb{Z}(G/N)$, which is denoted by $\tilde{\phi}_N$. Clearly it follows that $\mathrm{Ker}\tilde{\phi}_N = I(N)\mathbf{Z}G = \mathbf{Z}GI(N)$. Then we have the following:

LEMMA 2.6. *Let* G *be any group,* M *and* N *be normal subgroups of* G *with* $M \supseteqq N$. *Then for any* $n \geq 1$, *we have an isomorphism* $\tilde{\phi}_N^*$ *of additive groups induced by* $\tilde{\phi}_N$:

$$\tilde{\phi}_N^*: \frac{I^n(G) + I(M)\mathbf{Z}G}{I^{n+1}(G) + I(M)I(G) + I(N)\mathbf{Z}G} \to \frac{I^n(G/N) + I(M/N)\mathbf{Z}(G/N)}{I^{n+1}(G/N) + I(M/N)I(G/N)}.$$

Proof. For simplicity we put $A_n = I^n(G) + I(M)\mathbf{Z}G$ and $B_n = I^{n+1}(G) + I(M)I(G) + I(N)\mathbf{Z}G$. Then it follows that $\tilde{\phi}_N(A_n) = I^n(G/N) + I(M/N)\mathbf{Z}(G/N)$ and $\tilde{\phi}_N(B_n) = I^{n+1}(G/N) + I(M/N)I(G/N)$. Therefore $\tilde{\phi}_N$ induces an additive group homomorphism $\tilde{\phi}_N^* : A_n/B_n \to I^n(G/N) + I(M/N)\mathbf{Z}(G/N)/I^{n+1}(G/N) + I(M/N)I(G/N)$. To prove $\tilde{\phi}_N^*$ is isomorphic, we have to show that $\tilde{\phi}_N^*$ is injective, since $\tilde{\phi}_N^*$ is surjective. Let α be any element of A_n, and assume that $\tilde{\phi}_N(\alpha)$ is contained in $I^{n+1}(G/N) + I(M/N)I(G/N) = \tilde{\phi}_N(B_n)$. There exists an element β in B_n such that $\tilde{\phi}_N(\alpha) = \tilde{\phi}_N(\beta)$, and hence $\alpha - \beta$ is contained in $\mathrm{Ker}\tilde{\phi}_N = I(N)\mathbf{Z}G \subseteqq B_n$. Therefore α is contained in B_n. Thus $\tilde{\phi}_N$ is injective, and hence $\tilde{\phi}_N^*$ is isomorphic. □

3. The result of Hales and Passi

Hales and Passi [3] gave a criterion for a nilpotent group of class 2 to be the circle group of a nilpotent ring of index 3.

For a group G, we denote by $Z(G)$ the center of G. Hales and Passi [3, Theorem 5] showed the following:

THEOREM 3.1. *Let* G *be a nilpotent group of class* 2. *Then the following are equivalent:*

(a) G *is the circle group of a nilpotent ring of index* 3.

(b) G *has a normal subgroup* N *of* G *such that*

1) $D_2(G) \subseteqq N \subseteqq Z(G)$,

2) *the homomorphism* $i : N \to \dfrac{I^2(G) + I(N)\mathbf{Z}G}{I^3(G) + I(N)I(G)}$ *splits, where*

$i(x) = x - 1 + I^3(G) + I(N)I(G) \quad (x \in N)$.

REMARKS. i) If G is the circle group of a nilpotent ring of index 3, a splitting homomorphism ψ of i is given by $\psi(\alpha + I^3(G) + I(N)I(G)) = \rho_G(\alpha)$ (an element considered to be in N) $(\alpha \in I^2(G) + I(N)\mathbf{Z}G)$.

ii) Assume that G is a nilpotent group of class 2 satisfying the condition (b). Then the condition 1) implies that the series $G = H_1 \supsetneqq H_2 = N \supseteqq H_3 = 1$ is an N-series of G, and $I^2(G) + I(N)\mathbf{Z}G = \Lambda_2$ and $I^3(G) + I(N)I(G) = \Lambda_3$, where $I(G) = \Lambda_1 \supseteqq \Lambda_2 \supseteqq \Lambda_3 \supseteqq \cdots$ is the canonical filtration of $I(G)$ with respect to the

N-series $G = H_1 \supseteqq H_2 = N \supseteqq H_3 = 1$. Moreover if ψ is a splitting homomorphism of i, G is isomorphic to $(I(G)/L, \circ)$ with $\mathrm{Ker}\psi = L/\Lambda_3$ (cf. [3, Theorem 5]).

4. Finite nilpotent groups of class 3

We recall general circumstances on the canonical filtrations of the augmentation ideals of the finite groups before we consider those of finite nilpotent groups of class 3. Let G be any finite group, and $\mathbb{Z}G$ be the group ring of G over $\mathbb{Z}$. Further G has a finite N-series $\mathbb{H} : G = H_1 \supseteqq H_2 \supseteqq \cdots \supseteqq H_c \supseteqq H_{c+1} = 1$. The N-series $\mathbb{H}$ induces a weight function w on G: for any element x of G

$$w(x) = \begin{cases} k & (x \in H_k - H_{k+1}) \\ \infty & (x = 1) \end{cases} .$$

Define a family $\{\Lambda_k\}_{k=0}^{\infty}$ of ideals of $\mathbb{Z}G$ as follows. Λ_k is spanned over $\mathbb{Z}$ by all products $(g_1 - 1)(g_2 - 1) \ldots (g_s - 1)$ with $\Sigma_{i=1}^{s} w(g_i) \geqq k$. Then Λ_k is an ideal of $\mathbb{Z}G$ with $\Lambda_0 = \mathbb{Z}G$, $\Lambda_1 = I(G)$, $\Lambda_k \supseteqq I^k(G)$, and $\Lambda_i . \Lambda_j \subseteqq \Lambda_{i+j}$ for all $i, j \geqq 1$. If $x \neq 1$, define $O^*(x)$ to be the order of the coset $\bar{x} = xH_{w(x)+1}$ in $H_{w(x)}/H_{w(x)+1}$. Each quotient group H_i/H_{i+1} is a finite abelian group whose operation is denoted additively, and hence there exist elements $x_{i1}, x_{i2}, \ldots, x_{i\lambda(i)}$ such that any element $\bar{g} \in H_i/H_{i+1}$ can be uniquely written in the form

$$\bar{g} = \alpha(1)\bar{x}_{i1} = \alpha(2)\bar{x}_{i2} + \ldots + \alpha(\lambda(i))\bar{x}_{i\lambda(i)} , \qquad \bar{x}_{ij} = x_{ij} + H_{i+1} ,$$

where $0 \leqq \alpha(j) < O^*(x_{ij})$ for all j. Moreover we can choose x_{ij} $(1 \leqq j \leqq \lambda(i))$ so that $O^*(x_{ij})$ divides $O^*(x_{ij+1})$ $(1 \leqq j \leqq \lambda(i) - 1)$. Set $\Phi = \{x_{ij} \mid i = 1, 2, \ldots, c,\ j = 1, 2, \ldots, \lambda(i)\}$. Order Φ by setting $x_{ij} < x_{kl}$ if $i < k$ or $i = k$ and $j < l$. Then every element $g \in G$ can be uniquely written in the form

$$g = \prod_{i=1}^{c} x_{i1}^{\alpha_{i1}} x_{i2}^{\alpha_{i2}} \ldots x_{i\lambda(i)}^{\alpha_{i\lambda(i)}} , \qquad (\#)$$

for some integers α_{ij} with $0 \leqq \alpha_{ij} < O^*(x_{ij})$, $1 \leqq i \leqq c$, $1 \leqq j \leqq \lambda(i)$, where Π runs in order of increasing i from left to right. Put $m = \Sigma_{i=1}^{c} \lambda(i)$, and consider an m-sequence $\alpha = (\alpha_{11}, \ldots, \alpha_{1\lambda(1)}, \alpha_{21}, \ldots, \alpha_{2\lambda(2)}, \ldots, \alpha_{c1}, \ldots, \alpha_{c\lambda(c)})$ for non-negative integers α_{ij}. The set of all m-sequences is well ordered lexicographically. An m-sequence $\alpha = (\alpha_{ij})$ is basic if $0 \leqq \alpha_{ij} < O^*(x_{ij})$ for all i and j. The weight $W(\alpha)$ of an m-sequence $\alpha = (\alpha_{ij})$ is defined to be

$$W(\alpha) = \sum_{\substack{1 \leqq i \leqq c \\ 1 \leqq j \leqq \lambda(i)}} w(x_{ij})\alpha_{ij} = \sum_{\substack{1 \leqq i \leqq c \\ 1 \leqq j \leqq \lambda(i)}} i\alpha_{ij} .$$

Given an m-sequence $\alpha = (\alpha_{ij})$, we define the proper product $P(\alpha) \in \mathbb{Z}G$ by

$$P(\alpha) = \prod_{i=1}^{c} (x_{i1}-1)^{\alpha_{i1}}(x_{i2}-1)^{\alpha_{i2}} \cdots (x_{i\lambda(i)}-1)^{\alpha_{i\lambda(i)}}$$

where Π runs in order of increasing i from left to right. When α is basic, we call $P(\alpha)$ to be a basic product. If $W(\alpha) \geq k$, then $P(\alpha) \in \Lambda_k$.

From now on assume that G is a finite nilpotent group of class 3 with N-series $G = H_1 \supseteq H_2 \supseteq H_3 \supseteq H_4 = 1$. For simplicity we put $\lambda(1) = s$, $\lambda(2) = t$, $\lambda(3) = u$; $O^*(x_{1i}) = d(i)$ $(1 \leq i \leq s)$, $O^*(x_{2l}) = e(l)$ $(1 \leq l \leq t)$, $O^*(x_{3m}) = f(m)$ $(1 \leq m \leq u)$. Then $\mathbb{Z}$-free bases of Λ_2 and Λ_3 are well-known (cf. [9, Lemmas 3.2 and 3.3]):

LEMMA 4.1. *Λ_2 has a $\mathbb{Z}$-free basis consisting of*

(1) $(x_{1i}-1)^{d(i)}$, $1 \leq i \leq s$;

(2) $P(\alpha)$, α *basic*, $W(\alpha) \geq 2$.

LEMMA 4.2. *Λ_3 has a $\mathbb{Z}$-free basis consisting of*

(1) $(x_{1i}-1)^{d(l)}$, $d(i) \geq 3$, $1 \leq i \leq s$;

(2) $(x_{2l}-1)^{e(l)}$, $1 \leq l \leq t$;

(3) $d(i)(x_{1i}-1)(x_{1j}-1)$, $1 \leq i \leq j \leq s$;

(4) $P(\alpha)$, α *basic*, $W(\alpha) \geq 3$.

For $n = 2$ and 3, we can define the canonical homomorphism $i_n : H_n/H_{n+1} \to \Lambda_n/\Lambda_{n+1}$ given by $i_n(x + H_{n+1}) = x - 1 + \Lambda_{n+1}$ $(x \in H_n)$. As is well known, i_2 splits since $\Lambda_2/\Lambda_3 \cong H_2/H_3 \oplus \mathrm{Sp}^2(H_1/H_2)$ ([8, Theorem 6]), and hence there is the projection map which gives a splitting homomorphism of i_2. But we will here consider a more general splitting homomorphism of i_2. Define a map $\Psi_2 : \Lambda_2 \to H_2/H_3$ by

$$\Psi_2((x_{1i}-1)^{d(i)}) = 0 \qquad (d(i) \geq 3,\ 1 \leq i \leq s),$$

$$\Psi_2((x_{1i}-1)(x_{1j}-1)) = \sum_{l=1}^{t} a_l^{(ij)} \bar{x}_{2l} \qquad (1 \leq i \leq j \leq s),$$

$$\Psi_2(x_{2l}-1) = \bar{x}_{2l} \qquad (1 \leq l \leq t),$$

$$\Psi_2(P(\alpha)) = 0 \qquad (\alpha \text{ basic},\ W(\alpha) \geq 3)$$

for some integers $a_l^{(ij)}$ $(1 \leq i \leq j \leq s,\ 1 \leq l \leq t)$. By Lemma 4.2, $\Psi_2(\Lambda_3) = 0$ if and only if $d(i)a_l^{(ij)} \equiv 0 (\mathrm{mod}\ e(l))$ $(1 \leq i \leq j \leq s,\ 1 \leq l \leq t)$. In this case Ψ_2 induces a splitting homomorphism ψ_2 of i_2. Put $\mathrm{Ker}\psi_2 = L_2/\Lambda_3$. Then we have easily the following

LEMMA 4.3. *L_2 is an ideal of $I(G)$, and has a $\mathbb{Z}$-free basis consisting of*

(1) $(x_{1i}-1)^{d(i)}$, $d(i) \geq 3$, $1 \leq i \leq s$;

(2) $(x_{2l}-1)^{e(l)}$, $1 \leq l \leq t$;

(3) $(x_{1i}-1)(x_{1j}-1) - \sum_{l=1}^{t} a_l^{(ij)}(x_{2l}-1)$, $1 \leq i \leq j \leq s$;

(4) $P(\alpha)$, α *basic*, $W(\alpha) \geq 3$.

Here we consider some special elements in L_2 :

$$X_{ij} = (x_{1i}-1)(x_{1j}-1) - \sum_{l=1}^{t} a_l^{(ij)}(x_{2l}-1) - \sum_{m=1}^{u} \tilde{a}_m^{(ij)}(x_{3m}-1)$$

for some integers $\tilde{a}_m^{(ij)}$, which we will use below.

In general, the homomorphism i_3 does not necessarily split (cf. [8, Theorem 7]). We consider a map $\Psi_3 : \Lambda_3 \to H_3$ defined by

$$\left.\begin{array}{ll}
\Psi_3((x_{1i}-1)^{d(i)}) = 0 & (d(i) \geq 4 \text{ , } 1 \leq i \leq s) \text{ ,} \\
\Psi_3((x_{1i}-1)^{d(i)}) = \sum_{m=1}^{u} b_m^{(i)} x_{3m} & (d(i) = 3 \text{ , } 1 \leq i \leq s) \text{ ,} \\
\Psi_3((x_{2l}-1)^{e(l)}) = 0 & (1 \leq l \leq t) \text{ ,} \\
\Psi_3(d(i)(x_{1i}-1)(x_{1j}-1)) = \sum_{m=1}^{u} c_m^{(ij)} x_{3m} & (1 \leq i \leq j \leq s) \text{ ,} \\
\Psi_3((x_{1i}-1)(x_{1j}-1)(x_{1k}-1)) = \sum_{m=1}^{u} d_m^{(ijk)} x_{3m} & (\text{basic, } 1 \leq i \leq j \leq k \leq s) \text{ ,} \\
\Psi_3((x_{1i}-1)(x_{2l}-1)) = \sum_{m=1}^{u} e_m^{(il)} x_{3m} & (1 \leq i \leq s \text{ , } 1 \leq l \leq t) \text{ ,} \\
\Psi_3(x_{3m}-1) = x_{3m} & (1 \leq m \leq u) \text{ ,} \\
\Psi_3(P(\alpha)) = 0 & (\alpha \text{ basic, } W(\alpha) \geq 4) \text{ ,}
\end{array}\right\} \quad (**)$$

for some integers $b_m^{(i)}$ $(d(i) = 3 \text{ , } 1 \leq i \leq s, 1 \leq m \leq u)$, $c_m^{(ij)}$ $(1 \leq i \leq j \leq s, 1 \leq m \leq u)$, $d_m^{(ijk)}$ $(\text{basic, } 1 \leq i \leq j \leq k \leq s, 1 \leq m \leq u)$ and $e_m^{(il)}$ $(1 \leq i \leq s, 1 \leq l \leq t, 1 \leq m \leq u)$. Then we wish to determine a necessary and sufficient condition such that $\Psi_3(\Lambda_4) = 0$. To do this we put

$$\begin{array}{lll}
[x_{1i}, x_{1j}] \equiv \sum_{l=1}^{t} \alpha_l^{(ij)} x_2 & (\text{mod } H_3) & (1 \leq i \leq j \leq s) \text{ ;} \\
[x_{1i}, x_{2l}] = \sum_{m=1}^{u} \beta_m^{(il)} x_{3m} & (1 \leq i \leq s, \ 1 \leq l \leq t) \text{ ;} & \\
x_{1i}^{d(i)} \equiv \sum_{l=1}^{t} \gamma_l^{(i)} x_{2l} & (\text{mod } H_3) & (1 \leq i \leq s) \text{ ;} \\
x_{2l}^{e(l)} = \sum_{m=1}^{u} \delta_m^{(l)} x_{3m} & (1 \leq l \leq t) \text{ .} &
\end{array}$$

For two integers $a, b \geq 2$, we denote by (a, b) the greatest common divisor of a and b , and put

$$\langle a, b\rangle = \begin{cases} 1 & (b \geq 3) \\ 2 & (b = 2) \end{cases}.$$

Then we have the following:

LEMMA 4.4. $\Psi_3(\Lambda_4) = 0$ *if and only if, for any* m *with* $1 \leq m \leq u$, $f(m)$ *divide all the integers of the following types* (1) ~ (5):

$$(1)\quad \langle d(j), d(i)\rangle \left\{c_m^{(ij)} + \binom{d(i)}{2} d_m^{(iij)} - \sum_{l=1}^{t} \gamma_l^{(i)} e_m^{(jl)} + \sum_{l=1}^{t} \gamma_l^{(i)} \beta_m^{(jl)}\right\} \qquad (1 \leq i < j \leq s) ;$$

$$(2)\quad \langle d(i), d(j)\rangle \left\{\frac{d(j)}{d(i)} c_m^{(ij)} + \binom{d(j)}{2} d_m^{(ijj)} - \sum_{l=1}^{t} \gamma_l^{(j)} e_m^{(il)}\right\} \qquad (1 \leq i \leq j \leq s) ;$$

$$(3)\quad (d(i), e(l))e_m^{(il)} \qquad (1 \leq i \leq s , 1 \leq l \leq t) ;$$

$$(4)\quad d(i)d_m^{(ijk)} \qquad (\text{basic}, 1 \leq i \leq j \leq k \leq s) ;$$

$$(5)\quad d(i)b_m^{(i)} \qquad (d(i) = 3 , 1 \leq i \leq s) ,$$

where $d_m^{(iij)} = 0$ *if* $d(i) = 2$, *and* $d_m^{(ijj)} = 0$ *if* $d(j) = 2$.

Proof. A system of $\mathbb{Z}$-generators of Λ_4 is given in [8, Lemma 4] (cf. [9, Lemma 3.4]). We shall check it on each generator of Λ_4 . Ψ_3 maps clearly $(x_{1i}-1)^{d(i)}$ $(d(i) \geq 4, 1 \leq i \leq s)$ and $(x_{2l}-1)^{e(l)}$ $(1 \leq l \leq t)$ to zero. Since $(x_{3m}-1)^{f(m)} = -f(m)(x_{3m}-1) + \Sigma a_\alpha P(\alpha)$ for some basic α with $W(\alpha) \geq 6$, $\Psi_3((x_{3m}-1)^{f(m)}) = -f(m)x_{3m} = 0$ $(1 \leq m \leq u)$. We have for $1 \leq i < j \leq s$ with $d(i) \geq 3$,

$$\begin{aligned}(x_{1i}-1)^{d(i)}(x_{1j}-1) &= -d(i)(x_{1i}-1)(x_{1j}-1) - \binom{d(i)}{2}(x_{1i}-1)^2(x_{1j}-1) \\ &\quad + (x_{1j}-1)(x_{1i}^{d(i)}-1) + ([x_{1i}^{d(i)}, x_{1j}]-1) \\ &\quad + \sum_{m=1}^{u} a_m(x_{3m}-1)^{f(m)} + \Sigma a_\alpha P(\alpha) ,\end{aligned}$$

with some basic α with $W(\alpha) \geq 4$, and hence

$$\begin{aligned}\Psi_3((x_{1i}-1)^{d(i)}(x_{1j}-1)) &= -\sum_{m=1}^{u} c_m^{(ij)} x_{3m} - \binom{d(i)}{2} \sum_{m=1}^{u} d_m^{(iij)} x_{3m} \\ &\quad + \sum_{m=1}^{u}\sum_{l=1}^{t} \gamma_l^{(i)} e_m^{(jl)} x_{3m} - \sum_{m=1}^{u}\sum_{l=1}^{t} \gamma_l^{(i)} \beta_m^{(jl)} x_{3m} .\end{aligned}$$

Therefore $f(m)$ divides

$$c_m^{(ij)} + \binom{d(i)}{2} d_m^{(iij)} - \sum_{l=1}^{t} \gamma_l^{(i)} e_m^{(jl)} + \sum_{l=1}^{t} \gamma_l^{(i)} \beta_m^{(jl)} ,$$

which is an integer of type (1) with $d(i) \geqq 3$. Next we consider $(x_{1i}-1)(x_{1j}-1)^{d(j)}$ for $1 \leqq i \leqq j \leqq s$ with $d(j) \geqq 3$. Since

$$(x_{1i}-1)(x_{1j}-1)^{d(j)} = -d(j)(x_{1i}-1)(x_{1j}-1) - \binom{d(j)}{2}(x_{1i}-1)(x_{1j}-1)^2 + (x_{1i}-1)(x_{1j}^{d(j)}-1) + \Sigma a_\alpha P(\alpha) ,$$

for some basic α with $W(\alpha) \geqq 4$, we have

$$\Psi_3((x_{1i}-1)(x_{1j}-1)^{d(j)}) = -\frac{d(j)}{d(i)} \sum_{m=1}^{u} c_m^{(ij)} x_{3m} - \binom{d(j)}{2} \sum_{m=1}^{u} d_m^{(ijj)} x_{3m} + \sum_{m=1}^{u} \sum_{l=1}^{t} \gamma_l^{(j)} e_m^{(il)} x_{3m} ,$$

and hence $f(m)$ divides

$$\frac{d(j)}{d(i)} c_m^{(ij)} + \binom{d(j)}{2} d_m^{(ijj)} - \sum_{l=1}^{t} \gamma_l^{(j)} e_m^{(il)} ,$$

which is an integer of type (2) with $d(j) \geqq 3$. Clearly we have

$$\Psi_3((d(i), e(l))(x_{1i}-1)(x_{2l}-1)) = (d(i), e(l)) \sum_{m=1}^{u} e_m^{(il)} x_{3m} \quad (1 \leqq i \leqq s,\ 1 \leqq l \leqq t),$$

and hence $(d(i), e(l))e_m^{(il)} \equiv 0 \pmod{f(m)}$. Finally we consider $d(i)(x_{1i}-1)(x_{1j}-1)(x_{1k}-1)$ for $1 \leqq i \leqq j \leqq k \leqq s$. If $(x_{1i}-1)(x_{1j}-1)(x_{1k}-1)$ is basic, then

$$\Psi_3(d(i)(x_{1i}-1)(x_{1j}-1)(x_{1k}-1)) = d(i) \sum_{m=1}^{u} d^{(ijk)} x_{3m} ,$$

and hence $d(i)d_m^{(ijk)} \equiv 0 \pmod{f(m)}$. If $i = j < k$ with $d(i) = 2$, then

$$d(i)(x_{1i}-1)^2(x_{1k}-1) = -d(i)^2(x_{1i}-1)(x_{1k}-1) + d(i)(x_{1k}-1)(x_{1i}^{d(i)}-1) - d(i)([x_{1i}, x_{1i}^{d(i)}]-1) + \sum_{m=1}^{u} a_m(x_{3m}-1)^{f(m)} + \Sigma b_\alpha P(\alpha)$$

for some basic α with $W(\alpha) \geqq 4$, and hence

$$\Psi_3(d(i)(x_{1i}-1)^2(x_{1k}-1)) = -d(i) \sum_{m=1}^{u} c_m^{(ik)} x_{3m} + d(i) \sum_{m=1}^{u} \sum_{l=1}^{t} \gamma_l^{(i)} e_m^{(kl)} x_{3m} - d(i) \sum_{m=1}^{u} \sum_{l=1}^{t} \gamma_l^{(i)} \beta_m^{(kl)} x_{3m} .$$

Therefore $f(m)$ divides

$$d(i)\left\{c_m^{(ij)} - \sum_{l=1}^{t} \gamma_l^{(i)} e_m^{(jl)} + \sum_{l=1}^{t} \gamma_l^{(i)} \beta_m^{(jl)}\right\}$$

for $1 \leqq i < j \leqq s$ with $d(i) = 2$, which is an integer of type (1) with $d(i) = 2$, since $d_m^{(iij)} = 0$ in this case. If $i \leq j = k$ with $d(j) = 2$, then $d(i)(x_{1i}-1)(x_{1j}-1)^2 = -d(i)^2(x_{1i}-1)(x_{1j}-1) + d(i)(x_{1i}-1)(x_{1j}^{d(j)}-1)$ and hence

$$\Psi_3(d(i)(x_{1i}-1)(x_{1j}-1)^2 = -d(i) \sum_{m=1}^{u} c_m^{(ij)} x_{3m} + d(i) \sum_{m=1}^{u} \sum_{l=1}^{t} \gamma_l^{(j)} e_m^{(il)} x_{3m} .$$

Therefore $f(m)$ divides

$$d(i)\left\{c_m^{(ij)} - \sum_{l=1}^{t} \gamma_l^{(j)} e_m^{(il)}\right\} ,$$

which is an integer of type (2) with $d(j) = 2$, since $d_m^{(ijj)} = 0$ in this case. If $i = j = k$ with $d(i) = 3$, then

$$\Psi_3(d(i)(x_{1i}-1)^3) = d(i) \sum_{m=1}^{u} b_m^{(i)} x_{3m} ,$$

and hence $d(i)b_m^{(i)} \equiv 0 \pmod{f(m)}$. Thus we complete the proof of Lemma 4.4.

If Ψ_3 satisfies Lemma 4.4, then Ψ_3 induces a homomorphism $\psi_3 : \Lambda_3/\Lambda_4 \to H_3$ defined by $\psi_3(\alpha + \Lambda_4) = \Psi_3(\alpha)$ $(\alpha \in \Lambda_3)$, and clearly ψ_3 is a splitting homomorphism of i_3. Put $\mathrm{Ker}\psi_3 = L_3/\Lambda_4$. We have easily the following:

LEMMA 4.5. *L_3 is an ideal of $I(G)$, and has a system of $\mathbb{Z}$-generators consisting of*

(1) $(x_{1i}-1)^{d(i)}$, $\quad d(i) \geq 4$, $1 \leq i \leq s$;

(2) $(x_{1i}-1)^{d(i)} - \sum_{m=1}^{u} b_m^{(i)}(x_{3m}-1)$, $\quad d(i) = 3$, $1 \leq i \leq s$;

(3) $(x_{2l}-1)^{e(l)}$, $\quad 1 \leq l \leq t$;

(4) $(x_{3m}-1)^{f(m)}$, $\quad 1 \leq m \leq u$;

(5) $d(i)(x_{1i}-1)(x_{1j}-1) - \sum_{m=1}^{u} c_m^{(ij)}(x_{3m}-1)$, $\quad 1 \leq i \leq j \leq s$;

(6) $(x_{1i}-1)(x_{1j}-1)(x_{1k}-1) - \sum_{m=1}^{u} d_m^{(ijk)}(x_{3m}-1)$, $\quad$ basic, $1 \leq i \leq j \leq k \leq s$;

(7) $(x_{1i}-1)(x_{2l}-1) - \sum_{m=1}^{u} e_m^{(il)}(x_{3m}-1)$, $\quad 1 \leq i \leq s$, $1 \leq l \leq t$;

(8) $P(\alpha)$, $\quad \alpha$ basic, $W(\alpha) \geq 4$.

Here we consider the ideal generated by

$$X_{ij} = (x_{1i}-1)(x_{1j}-1) - \sum_{l=1}^{t} a_l^{(ij)}(x_{2l}-1) - \sum_{m=1}^{u} \tilde{a}_m^{(ij)}(x_{3m}-1) \quad (1 \leq i \leq j \leq s)$$

for some integers $\tilde{a}_m^{(ij)}$ $(1 \leq m \leq u)$ and L_3 in the ring $I(G)$, and we denote it by M, which is also an ideal in $\mathbb{Z}G$. We are now on the stage to state our main result. It is a generalization of Theorem 3.1 to the case of finite nilpotent groups of class 3 .

THEOREM 4.6. *Let G be a finite nilpotent group of class 3 . Then the following are equivalent:*

(a) *G is the circle group of a nilpotent ring of index 4 .*

(b) *G has an N-series: $G = H_1 \supseteqq H_2 \supseteqq H_3 \supseteqq H_4 = 1$ such that for $n = 2$ and 3 ,*

1) $D_n(G) \subseteqq H_n$,

2) *the homomorphism* $i_n : H_n/H_{n+1} \to \Lambda_n/\Lambda_{n+1}$ *splits* ,

3) *there exist integers* $a_l^{(ij)}$, $\tilde{a}_m^{(ij)}$ $(1 \leqq i \leqq j \leqq s,\ 1 \leqq l \leqq t,\ 1 \leqq m \leqq u)$ *such that* $H_3 \cap (1 + M) = 1$.

Proof. Let R be a nilpotent ring of index 4 , and $G = (R, \circ)$ be the circle group of R . Then we shall show the group G satisfies the condition (b). Put $H_n = (R^n, \circ)$ for any integer n with $1 \leqq n \leqq 4$, then H_n is a normal subgroup in G by Lemma 2.3, and $G = H_1 \supseteqq H_2 \supseteqq H_3 \supseteqq H_4 = 1$ is an N-series by Corollary 2.2. For $n = 2$ and 3 , $D_n(G) \subseteqq H_n$ by Corollary 2.5, which means 1). We consider the canonical ring homomorphism $\rho_G : I(G) \to R$. Then we have clearly $\rho_G(\Lambda_n) = R^n$ for $1 \leqq n \leqq 4$, and hence we can define a homomorphism $\psi_n : \Lambda_n/\Lambda_{n+1} \to H_n/H_{n+1}$ by $\psi_n(\alpha + \Lambda_{n+1}) = \rho_G(\alpha) + H_{n+1}$ $(\alpha \in \Lambda_n)$ for $n = 2$ and 3 . Then clearly $\psi_n \cdot i_n$ is the identity map on H_n/H_{n+1} , and hence the homomorphism $i_n : H_n/H_{n+1} \to \Lambda_n/\Lambda_{n+1}$ splits for $n = 2$ and 3 , which implies 2).

We distinguish the product of elements in R and that of elements in G by denoting

$$x^n = \underbrace{x.x.\ldots.x}_{n} \quad \text{and} \quad x^{(n)} = \underbrace{x \circ x \circ \ldots \circ x}_{n} ,$$

respectively. Put $\mathrm{Ker}\psi_n = L_n/\Lambda_{n+1}$ for $n = 2$ and 3 . Then L_2 is generated by the elements of types (1) ~ (4) in Lemma 4.3, where the integers $a_l^{(ij)}$ and $\tilde{a}_m^{(ij)}$ are determined by

$$x_{1i}x_{1j} = \sum_{l=1}^{t} (a_l^{(ij)})x_{2l} + \sum_{m=1}^{u} (\tilde{a}_m^{(ij)})x_{3m} , \qquad 1 \leqq i \leqq j \leqq s ,$$

in H_2 . Furthermore L_3 is generated by the elements of types (1) ~ (8) in Lemma 4.5. Here the integers $b_m^{(i)}$, $c_m^{(ij)}$, $d_m^{(ijk)}$ and $e_m^{(il)}$ are determined by

$$x_{1i}^{d(i)} = \sum_{m=1}^{u} (b_m^{(i)})x_{3m} , \qquad d(i) = 3 , \quad 1 \leqq i \leqq s ;$$

$$d(i)x_{1i}x_{1j} = \sum_{m=1}^{u} (c_m^{(ij)})x_{3m} , \qquad 1 \leqq i \leqq j \leqq s ;$$

$$x_{1i}x_{1j}x_{1k} = \sum_{m=1}^{u} (d_m^{(ijk)})x_{3m} , \qquad \text{basic}, \quad 1 \leqq i \leqq j \leqq k \leqq s ;$$

$$x_{1i}x_{2l} = \sum_{m=1}^{u} (e_m^{(il)})x_{3m} , \qquad 1 \leqq i \leqq s , \quad 1 \leqq l \leqq t ,$$

in H_3. Let M be the ideal in $I(G)$ generated by

$$X_{ij} = (x_{1i}-1)\circ(x_{1j}-1) - \sum_{l=1}^{t} a_l^{(ij)}\cdot(x_{2l}-1) - \sum_{m=1}^{u} \tilde{a}_m^{(ij)}\cdot(x_{3m}-1) \quad (1 \leqq i \leqq j \leqq s)$$

and L_3. Then it is easy to see, that $\rho_G(X_{ij}) = 0$ and the image of each $\mathbb{Z}$-generator of L_3 by ρ_G is zero. Thus we have $\rho_G(M) = 0$. Let x be any element of $H_3 \cap (1+M)$. Then $x-1$ is contained in M, and hence $x = \rho_G(x-1) \in \rho_G(M) = 0$. Therefore we get $H_3 \cap (1+M) = 1$, which implies 3). Thus we show that $G = (R, \circ)$ satisfies the condition (b).

Conversely we assume that a finite nilpotent group G of class 3 satisfies the condition (b). For $n = 2$ and 3, let $\psi_n : \Lambda_n/\Lambda_{n+1} \to H_n/H_{n+1}$ be a splitting homomorphism of i_n, and $\mathrm{Ker}\psi_n = L_n/\Lambda_{n+1}$. Put $G^* = G/H_3$ and $H_n^* = H_n/H_3$ for $n = 1$, 2 and 3. Then $G^* = H_1^* \supseteqq H_2^* \supseteqq H_3^* = 1$ is an N-series of G^*, and hence G^* is of class at most 2. Suppose G^* is of class 1, then $1 = (G^*)_2 = [G^*, G^*] = G_2H_3/H_3$ and hence $G_2 \subseteqq H_3$. Therefore we have $G_3 = [G, G_2] \subseteqq [H_1, H_3] \subseteqq H_4 = 1$, which is a contradiction. Thus G^* is of class 2. Let $I(G^*) = \Lambda_1^* \supseteqq \Lambda_2^* \supseteqq \Lambda_3^* \supseteqq \ldots$ be the canonical filtration of $I(G^*)$ with respect to the N-series $G^* = H_1^* \supseteqq H_2^* \supseteqq H_3^* = 1$. Then we have $\Lambda_n^* = \tilde{\phi}_{H_3}(\Lambda_n)$ for any $n \geqq 1$, and $\tilde{\phi}_{H_3}^* : \Lambda_n/\Lambda_{n+1} \cong \Lambda_n^*/\Lambda_{n+1}^*$ for $n = 1$, 2 and 3 by Lemma 2.6. Consider a homomorphism $i_2^* = \tilde{\phi}_{H_3}^* . i_2 : H_2^* \to \Lambda_2^*/\Lambda_3^*$, then i_2^* splits and its splitting homomorphism is given by $\psi_2^* = \psi_2(\phi_{H_3}^*)^{-1}$. If we denote $\mathrm{Ker}\psi_2^* = L_2^*/\Lambda_3^*$, then $L_2^* = \tilde{\phi}_{H_3}(L_2)$ and G^* is isomorphic to $(I(G^*)/L_2^*, \circ)$ by the remark below Theorem 3.1. Assume that L_2 and L_3 are realized by Lemma 4.3 and Lemma 4.5, respectively. Now we consider the ideal M of $I(G)$ generated by X_{ij} $(1 \leqq i \leqq j \leqq s)$ and L_3. We shall show H_3 is isomorphic to the additive group L_2/M. Let μ be a map from H_3 to L_2/M defined by $\mu(x) = x-1+M$ $(a \in H_3)$. Then μ is a homomorphism since $\Lambda_4 \subseteqq M$. The condition 3) implies that μ is injective. So we have to show μ is surjective. We will mark each element of types (1), (2), (3) and (4) of Lemma 4.3. The element of type I1) is congruent to

$$\sum_{m=1}^{u} b_m^{(i)}(x_{3m}-1) \quad \text{or} \quad 0$$

modulo M according to $d(i) = 3$ or $d(i) \geqq 4$. Clearly the element of type (2) is congruent to 0 modulo M. The element of type (3) is congruent to

$$\sum_{m=1}^{u} \tilde{a}_m^{(ij)}(x_{3m}-1)$$

modulo M. The elements of type (4) with $W(\alpha) = 3$ are only $(x_{1i}-1)(x_{1j}-1)(x_{1k}-1)$ (basic, $1 \leqq i \leqq j \leqq k \leqq s$), $(x_{1i}-1)(x_{2l}-1)$ $(1 \leqq i \leqq s,\ 1 \leqq l \leqq t)$ and $(x_{3m}-1)$ $(1 \leqq m \leqq u)$. These elements are congruent to

$$\sum_{m=1}^{u} d_m^{(ijk)}(x_{3m}-1), \qquad \sum_{m=1}^{u} e_m^{(il)}(x_{3m}-1)$$

and $(x_{3m}-1)$ modulo M, respectively. Further the elements of type (4) with $W(\alpha) \geqq 4$ are congruent to 0 modulo M. Thus any element of L_2/M is written by a $\mathbb{Z}$-linear combination of $x_{3m}-1$ $(1 \leqq m \leqq u)$ modulo M, and hence the homomorphism μ is surjective. Therefore H_3 is isomorphic to L_2/M by μ, and obviously $L_2/M \cong (L_2/M, \circ)$ since L_2/M is a nilpotent ring of index 2.

Now we note the ring $I(G)/M$ is a nilpotent ring of index at most 4, since $I^4(G) \subseteq \Lambda_4 \subseteq M$. We consider a map $\Psi : G \to (I(G)/M, \circ)$ defined by $\Psi(x) = x - 1 + M$ $(x \in G)$. Then the map Ψ is a group homomorphism from G to $(I(G)/M, \circ)$ and the restriction of Ψ to H_3 is equal to the above homomorphism μ. Hence we get the following commutative diagram:

$$\begin{array}{ccc} H_3 & \xrightarrow[\sim]{\mu} & (L_2/M, \circ) \\ \downarrow & & \downarrow \\ G & \xrightarrow{\Psi} & (I(G)/M, \circ) \\ \downarrow & & \downarrow \tilde{\phi}_{H_3} \\ G^* & \xrightarrow{\sim} & (I(G^*)/\tilde{\phi}_{H_3}(L_2), \circ) . \end{array}$$

Thus G is isomorphic to $(I(G)/M, \circ)$. Finally we shall show the ring $I(G)/M$ is just a nilpotent ring of index 4. Suppose that $(I(G)/M)^3 = 0$. Since $((I(G)/M)^3, \circ) \supseteq D_3(G) = G_3$ by Corollary 2.5, we have $G_3 = 1$, which is a contradiction, and hence $(I(G)/M)^3 \neq 0$. This completes the proof of the theorem. □

We determine both subsets $X_{ij}I(G)$ and $I(G)X_{ij}$ for $1 \leqq i \leqq j \leqq s$.

LEMMA 4.7. *Any element of the ideal of* $I(G)$ *generated by* X_{ij} $(1 \leqq i \leqq j \leqq s)$ *is,* modulo Λ_4, *a* $\mathbb{Z}$*-linear combination of the following elements:*

(1) $$(x_{1i}-1)(x_{1j}-1)(x_{1k}-1) - \sum_{l=1}^{t} a_l^{(jk)}(x_{1i}-1)(x_{2l}-1) ;$$

(2) $$(x_{1i}-1)([x_{1j}, x_{1k}]-1) - \sum_{l=1}^{t} a_l^{(jk)}(x_{1i}-1)(x_{2l}-1) + \sum_{l=1}^{t} a_l^{(ik)}(x_{1j}-1)(x_{2l}-1) - \sum_{l=1}^{t} a_l^{(ik)}([x_{1j}, x_{2l}]-1) ;$$

(3) $$(x_{1j}-1)([x_{1i}, x_{1k}]-1) + \sum_{l=1}^{t} a_l^{(jk)}(x_{1i}-1)(x_{2l}-1) - \sum_{l=1}^{t} a_l^{(ik)}(x_{1j}-1)(x_{2l}-1) - \sum_{l=1}^{t} a_l^{(jk)}([x_{1i}, x_{2l}]-1) ;$$

(4) $$(x_{1k}-1)([x_{1i}, x_{1j}]-1) - \sum_{l=1}^{t} a_l^{(jk)}(x_{1i}-1)(x_{2l}-1) + \sum_{l=1}^{t} a_l^{(ik)}(x_{1i}-1)(x_{2l}-1) + ([x_{1i}, x_{1j}, x_{1k}]-1) ;$$

(5) $$\sum_{l=1}^{t} a_l^{(jk)}(x_{1i}-1)(x_{2l}-1) - \sum_{l=1}^{t} a_l^{(ij)}(x_{1k}-1)(x_{2l}-1) + \sum_{l=1}^{t} a_l^{(ij)}([x_{1k}, x_{2l}]-1) ;$$

(6) $$\sum_{l=1}^{t} a_l^{(jk)}([x_{1i}, x_{2l}]-1) + \sum_{l=1}^{t} a_l^{(ik)}([x_{1j}, x_{2l}]-1) + \sum_{l=1}^{t} a_l^{(ij)}([x_{1k}, x_{2l}]-1) + ([x_{1i}, x_{1k}, x_{1j}]-1) ,$$

for all i , j *and* k *with* $1 \leq i \leq j \leq k \leq s$.

Proof. Put

$$X_{ij}^{*} = (x_{1i}-1)(x_{1j}-1) - \sum_{l=1}^{t} a_l^{(ij)}(x_{2l}-1) \qquad (1 \leq i \leq j \leq s) .$$

Since we can consider it modulo Λ_4 , we have to mark these elements $X_{ij}^{*}(x_{1k}-1)$ and $(x_{1k}-1)X_{ij}^{*}$ for $1 \leq i \leq j \leq s$ and $1 \leq k \leq s$. For given X_{ij}^{*} with $1 \leq i \leq j \leq s$ we consider $(x_{1k}-1)X_{ij}^{*}$ with $1 \leq k \leq i$. Then we have

$$(x_{1k}-1)X_{ij}^{*} = (x_{1k}-1)(x_{1i}-1)(x_{1j}-1) - \sum_{l=1}^{t} a_l^{(ij)}(x_{1k}-1)(x_{2l}-1) .$$

By renumbering $k \to i$, $i \to j$ and $j \to k$, it is reduced to the element of type (1). For given X_{ij}^{*} with $1 \leq i \leq j \leq s$, consider $X_{ij}^{*}(x_{1k}-1)$ with $i \leq k \leq j$. Then we have

$$\begin{aligned} X_{ij}^{*}(x_{1k}-1) &= (x_{1i}-1)(x_{1j}-1)(x_{1k}-1) - \sum_{l=1}^{t} a_l^{(ij)}(x_{2l}-1)(x_{1k}-1) \\ &\equiv (x_{1i}-1)\{(x_{1k}-1)(x_{1j}-1) + ([x_{1j}, x_{1k}]-1)\} \\ &\quad - \sum_{l=1}^{t} a_l^{(ij)}\{(x_{1k}-1)(x_{2l}-1) + ([x_{2l}, x_{1k}]-1)\} \qquad (\text{mod } \Lambda_4) \\ &\equiv (x_{1i}-1)(x_{1k}-1)(x_{1j}-1) - (x_{1i}-1)([x_{1k}, x_{1j}]-1) \\ &\quad - \sum_{l=1}^{t} a_l^{(ij)}(x_{1k}-1)(x_{2l}-1) + \sum_{l=1}^{t} a_l^{(ij)}([x_{1k}, x_{2l}]-1) \qquad (\text{mod } \Lambda_4) . \end{aligned}$$

By renumbering $i \to i$, $k \to j$ and $j \to k$, it is reduced to the following element:

$$(x_{1i}-1)(x_{1j}-1)(x_{1k}-1)-(x_{1i}-1)([x_{1j},\, x_{1k}]-1)-\sum_{l=1}^{t} a_l^{(ik)}(x_{1j}-1)(x_{2l}-1) + \sum_{l=1}^{t} a_l^{(ik)}([x_{1j},\, x_{2l}]-1) .$$

Therefore we have the element of type (2) by substituting (1) in the above. We can quite similarly get the elements of types (3) ~ (6). We omit detailed proofs. □

Assume that a group G has an N-series $G = H_1 \supsetneqq H_2 \supsetneqq H_3 \supsetneqq H_4 = 1$, and consider maps $\Psi_2 : \Lambda_2 \to H_2/H_3$ defined by (*) and $\Psi_3 : \Lambda_3 \to H_3$ defined by (**). Then Ψ_2 induces a splitting homomorphism ψ_2 of i_2 if and only if $d(i)a_l^{(ij)} \equiv 0 \pmod{e(l)}$ $(1 \leq i \leq j \leq s,\ 1 \leq l \leq t)$, and Ψ_3 induces a splitting homomorphism ψ_3 of i_3 if and only if integers $b_m^{(i)}$, $c_m^{(ij)}$, $d_m^{(ijk)}$ and $e_m^{(il)}$ satisfy Lemma 4.4. Now consider the ideal M of $I(G)$ generated by X_{ij} $(1 \leq i \leq j \leq s)$ and L_3 with $\mathrm{Ker}\psi_3 = L_3/\Lambda_4$. Then a system of $\mathbb{Z}$-generators of M are given by all elements of types (1) ~ (8) of Lemma 4.5 and all elements of types (1) ~ (6) of Lemma 4.7. We wish to determine the condition such that the ideal M satisfy 3) of (b) in Theorem 4.6. Assume that $H_3 \cap (1+M) = 1$. If the product $(x_{1i}-1)(x_{1j}-1)(x_{1k}-1)$ is basic, then by Lemmas 4.5 and 4.7, the element

$$\sum_{m=1}^{t} \Big(d_m^{(ijk)} - \sum_{l=1}^{t} a_l^{(jk)} e_m^{(il)}\Big)(x_{3m}-1)$$

is contained in M , and hence

$$d_m^{(ijk)} = \sum_{l=1}^{t} a_l^{(jk)} e_m^{(il)} . \tag{i}$$

If we take $i = j = k$ and $d(i) = 3$ in (1) of Lemma 4.7, the element

$$\sum_{l=1}^{u} \Big(b_m^{(i)} - \sum_{l=1}^{t} a_l^{(ii)} e_m^{(il)}\Big)(x_{3m}-1)$$

is contained in M , and hence

$$b_m^{(i)} = \sum_{l=1}^{t} a_l^{(ii)} e_m^{(il)} . \tag{ii}$$

Moreover by Lemma 4.5, the element

$$\sum_{m=1}^{u} c_m^{(ij)}(x_{3m}-1) - \sum_{l=1}^{t} d(i)a_l^{(ij)}(x_{2l}-1) = \sum_{m=1}^{u} \Big(c_m^{(ij)} - \sum_{l=1}^{t} \frac{d(i)a_l^{(ij)}}{e(l)}\, \delta_m^{(l)}\Big)(x_{3m}-1)$$

is contained in M , and hence

$$c_m^{(ij)} = \sum_{l=1}^{t} \frac{d(i)a_l^{(ij)}}{e(l)}\, \delta_m^{(l)} . \tag{iii}$$

Thus the integers $b_m^{(i)}$, $c_m^{(ij)}$ and $d_m^{(ijk)}$ are determined by integers $a_l^{(ij)}$, $e_m^{(il)}$ and $\delta_m^{(l)}$. Then we have the following:

PROPOSITION 4.8. *Let* $\Psi_2 : \Lambda_2 \to H_2/H_3$ *and* $\Psi_3 : \Lambda_3 \to H_3$ *be maps defined by* (*) *and* (**). *Then* Ψ_2 *and* Ψ_3 *induce splitting homomorphisms of* i_2 *and* i_3, *and* $H_3 \cap (1+M) = 1$, *if and only if*

(0) $d(i)a_l^{(ij)} \equiv 0 \pmod{e(l)}$, $1 \leq i \leq j \leq s,\ 1 \leq l \leq t,$

and for any m *with* $1 \leq m \leq u$, $f(m)$ *divides all the integers of the following types* (1) ~ (8):

(1) $(d(i),\ e(l))e_m^{(il)}$, $\qquad 1 \leq i \leq s,\ 1 \leq l \leq t;$

(2) $$\sum_{l=1}^{t}\left\{\frac{d(i)a_l^{(ij)}}{e(l)}\delta_m^{(l)} + \binom{d(i)}{2}a_l^{(ij)}e_m^{(il)} - \gamma_l^{(j)}\beta_m^{(jl)}\right\} + d(j)\tilde{a}_m^{(ij)},$$ $1 \leq i \leq j \leq s;$

(3) $$\sum_{l=1}^{t}\left\{\frac{d(j)a_l^{(ij)}}{e(l)}\delta_m^{(l)} + \binom{d(j)}{2}a_l^{(jj)}e_m^{(il)} - \gamma_l^{(j)}e_m^{(il)}\right\} + d(j)\tilde{a}_m^{(ij)},$$ $1 \leq i \leq j \leq s;$

(4) $$\sum_{l=1}^{t}\left(\alpha_l^{(jk)}e_m^{(il)} - a_l^{(jk)}e_m^{(il)} + a_l^{(ik)}e_m^{(jl)} - a_l^{(ik)}\beta_m^{(jl)}\right), \qquad 1 \leq i \leq j \leq k \leq s;$$

(5) $$\sum_{l=1}^{t}\left(\alpha_l^{(ik)}e_m^{(jl)} + a_l^{(jk)}e_m^{(il)} - a_l^{(ik)}e_m^{(jl)} - a_l^{(jk)}\beta_m^{(il)}\right), \qquad 1 \leq i \leq j \leq k \leq s;$$

(6) $$\sum_{l=1}^{t}\left(\alpha_l^{(ij)}e_m^{(kl)} - a_l^{(jk)}e_m^{(il)} + a_l^{(ik)}e_m^{(jl)} - \alpha_l^{(ij)}\beta_m^{(kl)}\right) \qquad 1 \leq i \leq j \leq k \leq s;$$

(7) $$\sum_{l=1}^{t}\left(a_l^{(jk)}e_m^{(il)} - a_l^{(ij)}e_m^{(kl)} + a_l^{(ij)}\beta_m^{(kl)}\right), \qquad 1 \leq i \leq j \leq k \leq s;$$

(8) $$\sum_{l=1}^{t}\left(a_l^{(jk)}\beta_m^{(il)} - a_l^{(ik)}\beta_m^{(jl)} + a_l^{(ij)}\beta_m^{(kl)} - \alpha_l^{(ik)}\beta_m^{(jl)}\right), \qquad 1 \leq i \leq j \leq k \leq s.$$

Proof. Suppose that Ψ_2 and Ψ_3 induce splitting homomorphisms of i_2 and i_3 respectively, and that $H_3 \cap (1+M) = 1$. Then $d(i)a_l^{(ij)} \equiv 0 \pmod{e(l)}$ ($1 \leq i \leq j \leq s,\ 1 \leq l \leq t$), and the integers $b_m^{(i)}$, $c_m^{(ij)}$, $d_m^{(ijk)}$ and $e_m^{(il)}$ satisfy Lemma 4.4, where the integers $b_m^{(i)}$, $c_m^{(ij)}$ and $d_m^{(ijk)}$ are given by (ii), (iii) and (i), respectively. The condition (1) of this proposition follows from (3) of Lemma 4.4. The condition (2) of this proposition with $d(i) \geq 3$ follows from (1) with $d(i) \geq 3$ in (1) of Lemma 4.4. If we take $i = j < k$ and $d(i) = 2$ in (1) of Lemma 4.7, the element

$$\sum_{m=1}^{u}\left(c_m^{(ij)} + \sum_{l=1}^{t} a_l^{(ij)}e_m^{(il)} - \sum_{l=1}^{t}\gamma_l^{(i)}e_m^{(jl)} + \sum_{l=1}^{t}\gamma_l^{(i)}\beta_m^{(jl)}\right)(x_{3m}-1)$$

is contained in M, and hence $f(m)$ divides

$$c_m^{(ij)} + \sum_{l=1}^{t} a_l^{(ij)} e_m^{(il)} - \sum_{l=1}^{t} \gamma_l^{(i)} e_m^{(jl)} + \sum_{l=1}^{t} \gamma_l^{(i)} \beta_m^{(jl)}$$

which is an integer of type (2) with $d(i) = 2$. Quite similarly the integers of type (3) are congruent to 0 modulo $f(m)$ by (2) of Lemma 4.4 and (1) of Lemma 4.7. The congruence of the integers of types (4) $\sim$ (8) are easily obtained by the conditions (2) $\sim$ (6) of Lemma 4.7.

Conversely suppose that the integers $a_l^{(ij)}$ and $e_m^{(il)}$ satisfy the conditions (0) $\sim$ (8) of this proposition. Then we define the integers $b_m^{(i)}$, $c_m^{(ij)}$ and $d_m^{(ijk)}$ by (ii), (iii) and (i), respectively. Then by condition (0), $\Psi_2(\Lambda_3) = 0$ and hence Ψ_2 induces a splitting homomorphism ψ_2 of i_2. Next we have easily the conditions (1), (2) and (3) of Lemma 4.4 follow from (2), (3) and (1) of this proposition, respectively, and hence (4) and (5) of Lemma 4.4 are satisfied by (i) and (ii). Therefore by Lemma 4.4, $\Psi_3(\Lambda_4) = 0$ and hence Ψ_3 induces a splitting homomorphism ψ_3 of i_3. Finally we shall show $H_3 \cap (1+M) = 1$. Let x be any element of $H_3 \cap (1+M)$. Then $x-1$ is contained in M, and hence $x-1$ is a $\mathbb{Z}$-linear combination of X_{ij}, the elements of types (1) $\sim$ (8) of Lemma 4.5, and the elements of types (1) $\sim$ (6) of Lemma 4.7. But we shall show that all the elements of types (1) $\sim$ (6) of Lemma 4.7 are $\mathbb{Z}$-linear combinations of the elements of types (2), (4) $\sim$ (8) of Lemma 4.5. We consider the element of type (1) of Lemma 4.7. If $(x_{1i}-1)(x_{1j}-1)(x_{1k}-1)$ is basic, then by (i),

$$(x_{1i}-1)(x_{1j}-1)(x_{1k}-1) - \sum_{l=1}^{t} a_l^{(jk)}(x_{1i}-1)(x_{2l}-1) = (x_{1i}-1)(x_{1j}-1)(x_{1k}-1)$$
$$- \sum_{m=1}^{u} d_m^{(ijk)}(x_{3m}-1) - \sum_{l=1}^{t} a_l^{(jk)}\left\{(x_{1i}-1)(x_{2l}-1) - \sum_{m=1}^{u} e_m^{(il)}(x_{3m}-1)\right\}.$$

If $i = j < k$ and $d(i) = 2$, then by (iii) and (2) of this proposition,

$$(x_{1i}-1)^2(x_{1j}-1) - \sum_{l=1}^{t} a_l^{(ij)}(x_{1i}-1)(x_{2l}-1) = -\left\{d(i)(x_{1i}-1)(x_{1j}-1)\right.$$
$$\left.- \sum_{m=1}^{u} c_m^{(ij)}(x_{3m}-1)\right\} + \sum_{l=1}^{t} \gamma_l^{(i)}\left\{(x_{1j}-1)(x_{2l}-1) - \sum_{m=1}^{u} e_m^{(jl)}(x_{3m}-1)\right\}$$
$$- \sum_{l=1}^{t} a_l^{(ij)}\left\{(x_{1i}-1)(x_{2l}-1) - \sum_{m=1}^{u} e_m^{(il)}(x_{3m}-1)\right\}$$
$$+ \sum_{m=1}^{u} a_m(x_{3m}-1)^{f(m)} + \sum_{\substack{\alpha \text{ basic} \\ W(\alpha)\geq 4}} b_\alpha P(\alpha).$$

Similarly we can see that if $1 \leq i \leq j \leq s$ and $d(j) = 2$,

$$(x_{1i}-1)(x_{1j}-1)^2 - \sum_{l=1}^{t} a_l^{(jj)}(x_{1i}-1)(x_{2l}-1)$$

is a $\mathbb{Z}$-linear combination of the elements of types (5), (7) and (8) by (iii) and (3) of this proposition. If $i = j = k$ and $d(i) = 3$, then

$$(x_{1i}-1)^{d(i)} - \sum_{l=1}^{t} a_l^{(ii)}(x_{1i}-1)(x_{2l}-1)$$

is a $\mathbb{Z}$-linear combination of the elements of types (2) and (7) by (ii). Thus all the elements of type (1) of Lemma 4.7 are $\mathbb{Z}$-linear combinations of the elements of types (2), (4) ~ (8) of Lemma 4.5. Using (4) ~ (7) of this proposition, by the same way, we can easily show that all the elements of types (2) ~ (5) of Lemma 4.7 are $\mathbb{Z}$-linear combinations of the elements of types (7) and (8) of Lemma 4.5, and the elements of type (6) of Lemma 4.7 are $\mathbb{Z}$-linear combinations of elements of type (8) of Lemma 4.5 by (8) of this proposition. Thus $x-1$ is a $\mathbb{Z}$-linear combination of X_{ij} and the elements of types (1) ~ (8) of Lemma 4.5. On the other hand, since $x \in H_3$, we can put

$$x = \sum_{m=1}^{u} \mu_m x_{3m} , \quad \mu_m \in \mathbb{Z} .$$

Therefore

$$x-1 = \sum_{m=1}^{\mu} \mu_m(x_{3m}-1) + \Sigma a_\alpha P(\alpha) ,$$

for some integers a_α with α basic, $W(\alpha) \geq 6$. Since $(x_{1i}-1)^{d(i)}$ $(1 \leq i \leq s)$ and $P(\alpha)$ (α basic, $W(\alpha) > 2$) are free over $\mathbb{Z}$, comparing the above presentation of $x-1$ with the presentation of the $\mathbb{Z}$-linear combination of X_{ij} and the elements of types (1) ~ (8) of Lemma 4.5, we can easily see that all the coefficients of the elements of types (1), (2), (6) and (7) are zero, and the coefficient of X_{ij} is equal to $-d(i)$ times that of the element of type (5). Therefore the term of the element of type (3) cancels with the term of $(x_{2l}-1)^{e(l)}$ which comes from the term of X_{ij} , and hence

$$\mu_m = f(m)\mu'_m , \quad \mu'_m \in \mathbb{Z} , \quad 1 \leq m \leq u .$$

Thus $x = 1$, and hence $H_3 \cap (1+M) = 1$. This completes the proof of this proposition. □

Kruse and Price [3, Corollary 1.6.9] proved that any p-group of order p^4 and of nilpotent class 3 can not be the circle group of any nilpotent ring of index 4 , but their method can not be available for p-groups of order p^5 . Now, as its application of the main theorem, we can show that there exist p-groups of order p^5 and of nilpotent class 3 which are the circle groups of nilpotent rings of index 4 , and moreover we can classify all groups of order p^5 and of nilpotent class 3 which are the circle groups of nilpotent rings of index 4 . For an odd prime p , we use a classification list of all groups of order p^5 in Bender [2]. If G is a group of order p^5 and of nilpotent class 3 , then G is one of types 4, 5, 6, 19, 20, 20_1,

21, 22, 23, 24, 37, 40, 41, 44, 45, 46, 47, 48, 49, 51, 53 and 54. Let G be the group of type 4 . Then G is generated by elements S_1, S_2, S_3, S_4 and S_5 with relations

$$S_1^p = S_2^p = S_3^p = S_4^p = S_5^p = 1 , \qquad Z(G) = \langle S_1, S_2 \rangle ,$$
$$[S_3, S_4] = 1 , \quad [S_3, S_5] = S_1 , \quad [S_4, S_5] = S_3 ,$$

where $Z(G)$ is the center of G . Putting $H_2 = \langle S_1, S_2, S_3 \rangle \supsetneqq H_3 = \langle S_1 \rangle$, $G = H_1 \supsetneqq H_2 \supsetneqq H_3 \supsetneqq H_4 = 1$ is an N-series of G . Taking $x_{11} = S_4$, $x_{12} = S_5$, $x_{21} = S_2$, $x_{22} = S_3$ and $x_{31} = S_1$, we have $d(1) = d(2) = e(1) = e(2) = f(1) = p$, $\alpha_1^{(12)} = 0$, $\alpha_2^{(12)} = 1$, $\beta_1^{(11)} = \beta_1^{(12)} = \beta_1^{(21)} = 0$, $\beta_1^{(22)} = -1$, and $\gamma_1^{(1)} = \gamma_2^{(1)} = \gamma_1^{(2)} = \gamma_2^{(2)} = \delta_1^{(1)} = \delta_1^{(2)} = 0$. Then if we put $a_1^{(11)} = a_1^{(22)} = a_2^{(11)} = a_2^{(22)} = e_1^{(12)} = e_1^{(11)} = 0$, $a_1^{(12)} = 1$, $a_2^{(12)} = a$, $\tilde{a}_1^{(11)} = \tilde{a}_1^{(12)} = \tilde{a}_1^{(22)} = 0$, $e_1^{(21)} = -a^2$ and $e_1^{(22)} = a-1$ with $2a-1 \equiv 0 \pmod p$, the integers $a_l^{(ij)}$ $(1 \leq i, j, l \leq 2)$ and $e_1^{(il)}$ $(1 \leq i, l \leq 2)$ satisfy the conditions (0) ~ (8) of Proposition 4.8, and hence the group G of type 4 is the circle group of a nilpotent ring of index 4 . Similarly we can show the groups of types 5, 19, 20_1, 21, 40, 41 and 54 are the circle groups of nilpotent rings of index 4 . Next let G be the group of type 6 . Then G is generated by elements S_1, S_3, S_4 and S_5 with relations

$$S_1^p = S_3^p = S_4^p = S_5^{p^2} = 1 , \qquad Z(G) = \langle S_1, S_5^p \rangle ,$$
$$[S_3, S_4] = 1 , \quad [S_3, S_5] = S_1 , \quad [S_4, S_5] = S_3 .$$

Then $G_2 = \langle S_1, S_3 \rangle \supsetneqq G_3 = \langle S_1 \rangle$. If $G = H_1 \supsetneqq H_2 \supsetneqq H_3 \supsetneqq H_4 = 1$ is any N-series of G , then one of the following three cases happens:

(1) $H_2 = G_2 \supset H_3 = G_3$

(2) $H_2 = \langle S_5^p, G_2 \rangle \supset H_3 = G_3$

(3) $H_2 = \langle S_5^p, G_2 \rangle \supset H_3 = \langle S_5^p, G_3 \rangle$.

For example, in case (2), we shall show that there do not exist integers $a_l^{(ij)}$, $\tilde{a}_m^{(ij)}$ and $e_m^{(il)}$ satisfying the conditions (0) ~ (8) of Proposition 4.8. Putting $x_{11} = S_4$, $x_{12} = S_5$, $x_{21} = S_5^p$, $x_{22} = S_3$ and $x_{31} = S_1$, we have $d(1) = d(2) = e(1) = e(2) = f(1) = p$, $\alpha_1^{(12)} = 0$, $\alpha_2^{(12)} = 1$, $\beta_1^{(11)} = \beta_1^{(12)} = \beta_1^{(21)} = 0$, $\beta_1^{(22)} = -1$, $\gamma_1^{(1)} = \gamma_2^{(1)} = 0$, $\gamma_1^{(2)} = 1$, $\gamma_2^{(2)} = \delta_1^{(1)} = \delta_1^{(2)} = 0$. Suppose there exist integers $a_l^{(il)}$ $(1 \leq i, j, l \leq 2)$, $a_1^{(ij)}$ $(1 \leq i \leq j \leq 2)$ and $e_1^{(il)}$ $(1 \leq i, l \leq 2)$ satisfying the conditions (0) ~ (8) of Proposition 4.8. We may neglect the integers $\tilde{a}_1^{(ij)}$ $(1 \leq i \leq j \leq 2)$ since $f(1) = p$. The conditions (3) ~ (6) imply $e_1^{(11)} \equiv e_1^{(21)} \equiv 0 \pmod p$, $a_2^{(22)} \equiv e_1^{(12)} \equiv a_2^{(12)}(e_1^{(22)}+1) \equiv 0 \pmod p$,

$(a_2^{(12)}-1)e_1^{(22)} \equiv 0 \pmod p$, and $(a_2^{(12)}+1)e_1^{(22)}+1 \equiv 0 \pmod p$, respectively. Therefore $a_2^{(12)} \equiv -e_1^{(22)} \pmod p$ and hence $e_1^{(22)}(e_1^{(22)}+1) \equiv (e_1^{(22)}-1)e_1^{(22)}-1 \equiv 0 \pmod p$. Thus $1 \equiv 0 \pmod p$ which is a contradiction. Thus there do not exist integers $a_l^{(ij)}$ $(1 \leq i, j, l \leq 2)$, $\tilde{a}_1^{(ij)}$ $(1 \leq i \leq j \leq 2)$ and $e_1^{(il)}$ $(1 \leq i, l \leq 2)$ satisfying (0) ~ (8) of Proposition 4.8. In cases (1) and (3), we can quite similarly show that there do not exist integers $a_l^{(ij)}$, $a_m^{(ij)}$ and $e_m^{(il)}$ satisfying (0) ~ (8) of Proposition 4.8, and hence the group G of type 6 is not the circle group of any nilpotent ring of index 4. Checking all other groups of the list by the same way as above, we have the following:

COROLLARY 4.9. *Let G be the group of order p^5 and of nilpotent class* 3 *with odd prime p. Then G is the circle group of a nilpotent ring of index* 4 *if and only if G is one of types* 4, 5, 19, 20_1, 41, 40, 41 *and* 54.

My graduate student K. Aoyama helped us to check this corollary.

References

[1] J.C. Ault and J.F. Watters, "Circle groups of nilpotent rings", *Amer. Math. Monthly* 80 (1973), pp. 48-52, MR47:5040.

[2] H.A. Bender, "A determination of the groups of order p^5", *Ann. of Math.* (2) 29 (1927), pp. 61-72, FdM53, p.105.

[3] A.W. Hales and I.B.S. Passi, "The second augmentation quotient of an integral group ring", *Arch. Math.* 31 (1978), pp. 259-265, MR81a:20010.

[4] R.L. Kruse and D.T. Price, *Nilpotent Rings* (Gordon and Breach, New York, London, Paris, 1969), MR42:1858.

[5] F. Röhl and J. Ullrich, "Über Komplettierungen und Lokalisierungen von Zirkelgruppen nilpotenter Ringe", *Arch. Math.* 37 (1981), pp. 300-305, MR83j:17002.

[6] R. Sandling, "Group rings of circle and unit groups", *Math. Z.* 140 (1974), pp. 195-202, MR52:3217.

[7] R. Sandling and K. Tahara, "Augmentation quotients of group rings and symmetric powers", *Math. Proc. Camb. Philos. Soc.* 85 (1979), pp. 247-252, MR80h:20014.

[8] K. Tahara, "On the structure of $Q_3(G)$ and the fourth dimension subgroups", *Japan. J. Math. (New Series)* 3 (1977), pp. 381-394, MR58:28157.

[9] K. Tahara, "The augmentation quotients of group rings and the fifth dimension subgroups", *J. Algebra* 71 (1981), pp. 141-173, MR83b:20007.

Department of Mathematics,
Aichi University of Education,
Igaya-cho, Kariya-shi,
Japan 448

APPENDIX A

Talks presented at "Groups-Korea 1983". An asterisk (*) marks those talks that are reproduced (in edited form) in these Proceedings.

(1) List of invited one-hour talks:

Gilbert Baumslag, **Finitely presented solvable groups etc.* (I, II)

Frank B. Cannonito, **Algorithmic problems for solvable groups* (I, II)

Mikhail Deza, *Permutation geometries and sharp graphs*

Narain D. Gupta, (I)*Free group rings*; *(II) *Extensions of groups and tree automorphisms*

Roger C. Lyndon, **Coset graphs* (I, II)

Jens L. Mennicke, **Discontinuous groups* (I, II)

Charles F. Miller III, (I) *The word problem for finitely generated solvable groups of finite rank*; (II) *Binding ties and subgroup theorems*

Michael F. Newman, **Metabelian groups of prime-power exponent*

Derek J.S. Robinson, *(I) *The word problem for finitely generated solvable groups of finite rank*; (II) *Solvable groups wity many polycyclic quotients*

Klaus W. Roggenkamp, **Isomorphisms and automorphisms of integral group rings*

(2) List of seminar talks:

Seymour Bachmuth, *Automorphisms of 2-generator metabelian groups*

Colin M. Campbell, **Presentations for simple groups*

Leo P. Comerford, Jr, *Equations over groups*

C. Kanta Gupta, *Subgroups of free groups – characteristic vs fully invariant*

David L. Johnson, **Circular braids*

Toru Maeda, *Lower central series of link groups*

Horace Y. Mochizuki, *Automorphisms of finitely generated metabelian groups*

Bernhard H. Neumann, **Commutative quandles*

Cheryl E. Praeger, **Finite simple groups, permutation groups, and symmetric graphs*

Stephen J. Pride, **The concept of largeness in group theory*

Akbar H. Rhemtulla, **Isolators in solvable groups*

Frank Röhl, **Induced automorphisms of group rings*

Ken-Ichi Tahara, **On the circular groups of finite nilpotent rings*

(3) List of talks given to the graduate students:

Gilbert Baumslag, (I) *Spheres, good groups, and nice functions*; (II) *The integral homology groups of finitely presented groups*

Frank B. Cannonito, *Group theory*

Colin M. Campbell, (I) *Graph and groups*; (II) *Coset ennumerations*; (III) *Fibonacci and Lucas numbers*

Kean Lee, *Group rings and cohomology of groups* (I, II)

Roger C. Lyndon, *Geometric methods in combinatorial group theory: An introduction* (I, II)

Jens L. Mennicke, *Discontinuous groups*

Bernhard H. Neumann, *Almost finite and almost abelian groups*

Michael F. Newman, **The composition problem for finite groups*

Sung-An Park, *A characterization of simple groups* (I, II)

APPENDIX B

List of Participants

Miss Kyung-Won An, Inha University
Mr Sang-Yook An, Yonsei University
Professor Seymour Bachmuth, University of California, Santa Barbara, USA
Professor Young-Bai Baik, Hyosung Women's University
Mr Young-Gil Baik, Busan National Fisheries University
Professor Gilbert Baumslag, City College of the City University of New York, USA
Professor Robert G. Burns, York University, Canada
Dr Colin M. Campbell, University of St Andrews, Scotland, UK
Professor Frank B. Cannonito, University of California, Irvine, USA
Mr Tae-Young, Choi, Keimyung University
Mr Jang-Ho Chun, Youngnam University
Professor Leo P. Comerford, Jr, University of Wisconsin, Parkside, USA
Mr Hean-Gi Dan, Chungbuk National University
Professor Mikhail Deza, Université Paris VII, France
Dr John R.J. Groves, University of Melbourne, Australia
Professor C. Kanta Gupta, University of Manitoba, Canada
Professor Narain D. Gupta, University of Manitoba, Canada
Professor Ki-Sik Ha, Busan National University
Mr Sang-Jung Ha, Busan National University
Dr Woo-Chorl Hong, Busan National University
Professor Verena Huber-Dyson, University of Calgary, Canada
Mr Tae-Young Huh, Busan National University
Mr Chan Huh, Busan National University
Professor Won Hu, Busan National University
Mr Tae-Hoon Hyun, Korea Military Academy
Miss In-Oak Hwang, Hyosung Women's University
Mr In-Kyu Jeong, Jeonju University
Mr Jong-Woo Jeong, Jeonbug National University
Miss Hyon-Joo Ji, Kyungpook National University
Dr David L. Johnson, University of Nottingham, England, UK, and Busan National University
Professor Ann-Chi Kim, Busan National University
Dr Chul-On Kim, Busan National University
Mr Dong-Hyeon Kim, Kongju Teacher's College
Mr Eun-Sub Kim, Kyungpook National University
Mr Hoe-Dong Kim, Kyungpook National University
Mr Hae-Gyu Kim, Kyungpook National University
Mr Hong-Goo Kim, Keimyung University
Mr Hoi-Ill Kim, Chungbuk National University
Miss Hae-Kyoung Kim, Kyungpook National University
Mr Jae-Kyoung Kim, Korea University
Professor Jong-Sik Kim, Seoul National University
Miss Kyoung-Rahan Kim, Busan National University
Miss Nang-Kum Kim, Busan National University
Mr Pan-Soo Kim, Busan National University
Professor Kyoung-Soo Kim, Busan National University
Mr Sang-Bum Kim, Kyungpook National University
Mr Jin-Jo Kwak, Kyungpook National University
Mr Gi-Ho Kwon, Jeonju University
Mr Jung-Yul Kwon, Keimyung University

Miss Youn-Ok Kwon, Kyungpook National University
Professor Chung-Gul Lee, Busan National University
Professor Kyung-Kyi Lee, Busan National University
Miss Jong-Suk Lee, Youngnam University
Professor Kean Lee, Jeonbug National University
Mr Kyoung-Wha Lee, Youngnam University
Professor Roger C. Lyndon, University of Michigan, USA
Dr Toru Maeda, University of Toronto, Canada
Professor Jens L. Mennicke, Universität Bielefeld, Federal Republic of Germany
Professor Charles F. Miller III, University of Melbourne, Australia
Professor Horace Y. Mochizuki, University of California, Santa Barbara, USA
Professor Bernhard H. Neumann, Australian National University, Australia
Dr Michael F. Newman, Australian National University, Australia
Miss Bog-Ryun Oh, Kyoungnam University
Mr Dum-Jil Park, Keimyung University
Mr Dae-Yeon Park, Korea University
Mr Gun Park, Busan National University
Dr Jae-Keol Park, Busan National University
Dr Jong-Yeoul Park, Busan National University
Professor Sung-An Park, Sogang University
Professor Sehie Park, Seoul National University
Miss Sung-Hee Park, Hyosung Women's University
Professor Sang-Gyu Park, Inha University
Mr Young-Ho Park, Yonsei University
Mr Young-Gou Park, Youngnam University
Professor Young-Soo Park, Kyungpook National University
Professor Young-Sik Park, Busan National University
Professor Cheryl E. Praeger, University of Western Australia, Australia
Dr Stephen J. Pride, University of Glasgow, UK
Professor Akbar H. Rhemtulla, University of Alberta, Canada
Mr Seog-Hoon Rim, Kyungpook National University
Professor Derek J.S. Robinson, University of Illinois at Urbana Champaign, USA
Professor Klaus W. Roggenkamp, Universität Stuttgart, Federal Republic of Germany
Dr Frank Röhl, Universität Hamburg, Federal Republic of Germany
Professor Gerhard Rosenberger, Universität Dortmund, Federal Republic of Germany
Mr Gyeung-Sik Seo, Jeonbug National University
Professor Tae-Young Seo, Busan National University
Mr Joon-Sang Shin, Sogang University
Mr Jin-Sik Shin, Chungbuk National University
Mr Joon-Young Shin, Busan National University
Mr Hyo-Sup Sim, Busan National University
Miss Kyung-Sun Sohn, Kyungpook National University
Professor Noon-Gu Sohn, Kyungpook National University
Mr Moo-Young Sohn, Kyungpook National University
Dr Tai-Sung Song, Busan National University
Professor Ken-Ichi Tahara, Aichi University of Education, Japan
Professor Jang-Il Um, Busan National University
Professor Malcolm J. Wicks, National University of Singapore, Singapore
Miss Soh-Mee Yoon, Kyoungnam University
Miss Kyong-Suk You, Kyungpook National University

Vol. 952: Combinatorial Mathematics IX. Proceedings, 1981. Edited by E. Billington, S. Oates-Williams, and A.P. Street. XI, 443 pages. 1982.

Vol. 953: Iterative Solution of Nonlinear Systems of Equations. Proceedings, 1982. Edited by R. Ansorge, Th. Meis, and W. Törnig. VII, 202 pages. 1982.

Vol. 954: S.G. Pandit, S.G. Deo, Differential Systems Involving Impulses. VII, 102 pages. 1982.

Vol. 955: G. Gierz, Bundles of Topological Vector Spaces and Their Duality. IV, 296 pages. 1982.

Vol. 956: Group Actions and Vector Fields. Proceedings, 1981. Edited by J.B. Carrell. V, 144 pages. 1982.

Vol. 957: Differential Equations. Proceedings, 1981. Edited by D.G. de Figueiredo. VIII, 301 pages. 1982.

Vol. 958: F.R. Beyl, J. Tappe, Group Extensions, Representations, and the Schur Multiplicator. IV, 278 pages. 1982.

Vol. 959: Géométrie Algébrique Réelle et Formes Quadratiques, Proceedings, 1981. Edité par J.-L. Colliot-Thélène, M. Coste, L. Mahé, et M.-F. Roy. X, 458 pages. 1982.

Vol. 960: Multigrid Methods. Proceedings, 1981. Edited by W. Hackbusch and U. Trottenberg. VII, 652 pages. 1982.

Vol. 961: Algebraic Geometry. Proceedings, 1981. Edited by J.M. Aroca, R. Buchweitz, M. Giusti, and M. Merle. X, 500 pages. 1982.

Vol. 962: Category Theory. Proceedings, 1981. Edited by K.H. Kamps D. Pumplün, and W. Tholen, XV, 322 pages. 1982.

Vol. 963: R. Nottrot, Optimal Processes on Manifolds. VI, 124 pages. 1982.

Vol. 964: Ordinary and Partial Differential Equations. Proceedings, 1982. Edited by W.N. Everitt and B.D. Sleeman. XVIII, 726 pages. 1982.

Vol. 965: Topics in Numerical Analysis. Proceedings, 1981. Edited by P.R. Turner. IX, 202 pages. 1982.

Vol. 966: Algebraic K-Theory. Proceedings, 1980, Part I. Edited by R.K. Dennis. VIII, 407 pages. 1982.

Vol. 967: Algebraic K-Theory. Proceedings, 1980. Part II. VIII, 409 pages. 1982.

Vol. 968: Numerical Integration of Differential Equations and Large Linear Systems. Proceedings, 1980. Edited by J. Hinze. VI, 412 pages. 1982.

Vol. 969: Combinatorial Theory. Proceedings, 1982. Edited by D. Jungnickel and K. Vedder. V, 326 pages. 1982.

Vol. 970: Twistor Geometry and Non-Linear Systems. Proceedings, 1980. Edited by H.-D. Doebner and T.D. Palev. V, 216 pages. 1982.

Vol. 971: Kleinian Groups and Related Topics. Proceedings, 1981. Edited by D.M. Gallo and R.M. Porter. V, 117 pages. 1983.

Vol. 972: Nonlinear Filtering and Stochastic Control. Proceedings, 1981. Edited by S.K. Mitter and A. Moro. VIII, 297 pages. 1983.

Vol. 973: Matrix Pencils. Proceedings, 1982. Edited by B. Kågström and A. Ruhe. XI, 293 pages. 1983.

Vol. 974: A. Draux, Polynômes Orthogonaux Formels – Applications. VI, 625 pages. 1983.

Vol. 975: Radical Banach Algebras and Automatic Continuity. Proceedings, 1981. Edited by J.M. Bachar, W.G. Bade, P.C. Curtis Jr., H.G. Dales and M.P. Thomas. VIII, 470 pages. 1983.

Vol. 976: X. Fernique, P.W. Millar, D.W. Stroock, M. Weber, Ecole d'Eté de Probabilités de Saint-Flour XI – 1981. Edited by P.L. Hennequin. XI, 465 pages. 1983.

Vol. 977: T. Parthasarathy, On Global Univalence Theorems. VIII, 106 pages. 1983.

Vol. 978: J. Ławrynowicz, J. Krzyż, Quasiconformal Mappings in the Plane. VI, 177 pages. 1983.

Vol. 979: Mathematical Theories of Optimization. Proceedings, 1981. Edited by J.P. Cecconi and T. Zolezzi. V, 268 pages. 1983.

Vol. 980: L. Breen. Fonctions thêta et théorème du cube. XIII, 115 pages. 1983.

Vol. 981: Value Distribution Theory. Proceedings, 1981. Edited by I. Laine and S. Rickman. VIII, 245 pages. 1983.

Vol. 982: Stability Problems for Stochastic Models. Proceedings, 1982. Edited by V. V. Kalashnikov and V. M. Zolotarev. XVII, 295 pages. 1983.

Vol. 983: Nonstandard Analysis-Recent Developments. Edited by A.E. Hurd. V, 213 pages. 1983.

Vol. 984: A. Bove, J. E. Lewis, C. Parenti, Propagation of Singularities for Fuchsian Operators. IV, 161 pages. 1983.

Vol. 985: Asymptotic Analysis II. Edited by F. Verhulst. VI, 497 pages. 1983.

Vol. 986: Séminaire de Probabilités XVII 1981/82. Proceedings. Edited by J. Azéma and M. Yor. V, 512 pages. 1983.

Vol. 987: C. J. Bushnell, A. Fröhlich, Gauss Sums and p-adic Division Algebras. XI, 187 pages. 1983.

Vol. 988: J. Schwermer, Kohomologie arithmetisch definierter Gruppen und Eisensteinreihen. III, 170 pages. 1983.

Vol. 989: A. B. Mingarelli, Volterra-Stieltjes Integral Equations and Generalized Ordinary Differential Expressions. XIV, 318 pages. 1983.

Vol. 990: Probability in Banach Spaces IV. Proceedings, 1982. Edited by A. Beck and K. Jacobs. V, 234 pages. 1983.

Vol. 991: Banach Space Theory and its Applications. Proceedings, 1981. Edited by A. Pietsch, N. Popa and I. Singer. X, 302 pages. 1983.

Vol. 992: Harmonic Analysis, Proceedings, 1982. Edited by G. Mauceri, F. Ricci and G. Weiss. X, 449 pages. 1983.

Vol. 993: R. D. Bourgin, Geometric Aspects of Convex Sets with the Radon-Nikodým Property. XII, 474 pages. 1983.

Vol. 994: J.-L. Journé, Calderón-Zygmund Operators, Pseudo-Differential Operators and the Cauchy Integral of Calderón. VI, 129 pages. 1983.

Vol. 995: Banach Spaces, Harmonic Analysis, and Probability Theory. Proceedings, 1980–1981. Edited by R.C. Blei and S.J. Sidney. V, 173 pages. 1983.

Vol. 996: Invariant Theory. Proceedings, 1982. Edited by F. Gherardelli. V, 159 pages. 1983.

Vol. 997: Algebraic Geometry – Open Problems. Edited by C. Ciliberto, F. Ghione and F. Orecchia. VIII, 411 pages. 1983.

Vol. 998: Recent Developments in the Algebraic, Analytical, and Topological Theory of Semigroups. Proceedings, 1981. Edited by K.H. Hofmann, H. Jürgensen and H. J. Weinert. VI, 486 pages. 1983.

Vol. 999: C. Preston, Iterates of Maps on an Interval. VII, 205 pages. 1983.

Vol. 1000: H. Hopf, Differential Geometry in the Large, VII, 184 pages. 1983.

Vol. 1001: D.A. Hejhal, The Selberg Trace Formula for PSL(2, IR). Volume 2. VIII, 806 pages. 1983.

Vol. 1002: A. Edrei, E.B. Saff, R.S. Varga, Zeros of Sections of Power Series. VIII, 115 pages. 1983.

Vol. 1003: J. Schmets, Spaces of Vector-Valued Continuous Functions. VI, 117 pages. 1983.

Vol. 1004: Universal Algebra and Lattice Theory. Proceedings, 1982. Edited by R. S. Freese and O.C. Garcia. VI, 308 pages. 1983.

Vol. 1005: Numerical Methods. Proceedings, 1982. Edited by V. Pereyra and A. Reinoza. V, 296 pages. 1983.

Vol. 1006: Abelian Group Theory. Proceedings, 1982/83. Edited by R. Göbel, L. Lady and A. Mader. XVI, 771 pages. 1983.

Vol. 1007: Geometric Dynamics. Proceedings, 1981. Edited by J. Palis Jr. IX, 827 pages. 1983.

Vol. 1008: Algebraic Geometry. Proceedings, 1981. Edited by J. Dolgachev. V, 138 pages. 1983.

Vol. 1009: T. A. Chapman, Controlled Simple Homotopy Theory and Applications. III, 94 pages. 1983.

Vol. 1010: J.-E. Dies, Chaînes de Markov sur les permutations. IX, 226 pages. 1983.

Vol. 1011: J. M. Sigal. Scattering Theory for Many-Body Quantum Mechanical Systems. IV, 132 pages. 1983.

Vol. 1012: S. Kantorovitz, Spectral Theory of Banach Space Operators. V, 179 pages. 1983.

Vol. 1013: Complex Analysis – Fifth Romanian-Finnish Seminar. Part 1. Proceedings, 1981. Edited by C. Andreian Cazacu, N. Boboc, M. Jurchescu and I. Suciu. XX, 393 pages. 1983.

Vol. 1014: Complex Analysis – Fifth Romanian-Finnish Seminar. Part 2. Proceedings, 1981. Edited by C. Andreian Cazacu, N. Boboc, M. Jurchescu and I. Suciu. XX, 334 pages. 1983.

Vol. 1015: Equations différentielles et systèmes de Pfaff dans le champ complexe – II. Seminar. Edited by R. Gérard et J. P. Ramis. V, 411 pages. 1983.

Vol. 1016: Algebraic Geometry. Proceedings, 1982. Edited by M. Raynaud and T. Shioda. VIII, 528 pages. 1983.

Vol. 1017: Equadiff 82. Proceedings, 1982. Edited by H. W. Knobloch and K. Schmitt. XXIII, 666 pages. 1983.

Vol. 1018: Graph Theory, Łagów 1981. Proceedings, 1981. Edited by M. Borowiecki, J. W. Kennedy and M. M. Sysło. X, 289 pages. 1983.

Vol. 1019: Cabal Seminar 79–81. Proceedings, 1979–81. Edited by A. S. Kechris, D. A. Martin and Y. N. Moschovakis. V, 284 pages. 1983.

Vol. 1020: Non Commutative Harmonic Analysis and Lie Groups. Proceedings, 1982. Edited by J. Carmona and M. Vergne. V, 187 pages. 1983.

Vol. 1021: Probability Theory and Mathematical Statistics. Proceedings, 1982. Edited by K. Itô and J.V. Prokhorov. VIII, 747 pages. 1983.

Vol. 1022: G. Gentili, S. Salamon and J.-P. Vigué. Geometry Seminar "Luigi Bianchi", 1982. Edited by E. Vesentini. VI, 177 pages. 1983.

Vol. 1023: S. McAdam, Asymptotic Prime Divisors. IX, 118 pages. 1983.

Vol. 1024: Lie Group Representations I. Proceedings, 1982–1983. Edited by R. Herb, R. Lipsman and J. Rosenberg. IX, 369 pages. 1983.

Vol. 1025: D. Tanré, Homotopie Rationnelle: Modèles de Chen, Quillen, Sullivan. X, 211 pages. 1983.

Vol. 1026: W. Plesken, Group Rings of Finite Groups Over p-adic Integers. V, 151 pages. 1983.

Vol. 1027: M. Hasumi, Hardy Classes on Infinitely Connected Riemann Surfaces. XII, 280 pages. 1983.

Vol. 1028: Séminaire d'Analyse P. Lelong – P. Dolbeault – H. Skoda. Années 1981/1983. Edité par P. Lelong, P. Dolbeault et H. Skoda. VIII, 328 pages. 1983.

Vol. 1029: Séminaire d'Algèbre Paul Dubreil et Marie-Paule Malliavin. Proceedings, 1982. Edité par M.-P. Malliavin. V, 339 pages. 1983.

Vol. 1030: U. Christian, Selberg's Zeta-, L-, and Eisensteinseries. XII, 196 pages. 1983.

Vol. 1031: Dynamics and Processes. Proceedings, 1981. Edited by Ph. Blanchard and L. Streit. IX, 213 pages. 1983.

Vol. 1032: Ordinary Differential Equations and Operators. Proceedings, 1982. Edited by W. N. Everitt and R. T. Lewis. XV, 521 pages. 1983.

Vol. 1033: Measure Theory and its Applications. Proceedings, 1982. Edited by J. M. Belley, J. Dubois and P. Morales. XV, 317 pages. 1983.

Vol. 1034: J. Musielak, Orlicz Spaces and Modular Spaces. V, 222 pages. 1983.

Vol. 1035: The Mathematics and Physics of Disordered Media. Proceedings, 1983. Edited by B. D. Hughes and B. W. Ninham. VII, 432 pages. 1983.

Vol. 1036: Combinatorial Mathematics X. Proceedings, 1982. Edited by L. R. A. Casse. XI, 419 pages. 1983.

Vol. 1037: Non-linear Partial Differential Operators and Quantization Procedures. Proceedings, 1981. Edited by S. I. Andersson and H.-D. Doebner. VII, 334 pages. 1983.

Vol. 1038: F. Borceux, G. Van den Bossche, Algebra in a Localic Topos with Applications to Ring Theory. IX, 240 pages. 1983.

Vol. 1039: Analytic Functions, Błażejewko 1982. Proceedings. Edited by J. Ławrynowicz. X, 494 pages. 1983

Vol. 1040: A. Good, Local Analysis of Selberg's Trace Formula. III, 128 pages. 1983.

Vol. 1041: Lie Group Representations II. Proceedings 1982–1983. Edited by R. Herb, S. Kudla, R. Lipsman and J. Rosenberg. IX, 340 pages. 1984.

Vol. 1042: A. Gut, K. D. Schmidt, Amarts and Set Function Processes. III, 258 pages. 1983.

Vol. 1043: Linear and Complex Analysis Problem Book. Edited by V. P. Havin, S. V. Hruščëv and N. K. Nikol'skii. XVIII, 721 pages. 1984.

Vol. 1044: E. Gekeler, Discretization Methods for Stable Initial Value Problems. VIII, 201 pages. 1984.

Vol. 1045: Differential Geometry. Proceedings, 1982. Edited by A. M. Naveira. VIII, 194 pages. 1984.

Vol. 1046: Algebraic K-Theory, Number Theory, Geometry and Analysis. Proceedings, 1982. Edited by A. Bak. IX, 464 pages. 1984.

Vol. 1047: Fluid Dynamics. Seminar, 1982. Edited by H. Beirão da Veiga. VII, 193 pages. 1984.

Vol. 1048: Kinetic Theories and the Boltzmann Equation. Seminar, 1981. Edited by C. Cercignani. VII, 248 pages. 1984.

Vol. 1049: B. Iochum, Cônes autopolaires et algèbres de Jordan. VI, 247 pages. 1984.

Vol. 1050: A. Prestel, P. Roquette, Formally p-adic Fields. V, 167 pages. 1984.

Vol. 1051: Algebraic Topology, Aarhus 1982. Proceedings. Edited by I. Madsen and B. Oliver. X, 665 pages. 1984.

Vol. 1052: Number Theory. Seminar, 1982. Edited by D. V. Chudnovsky, G. V. Chudnovsky, H. Cohn and M. B. Nathanson. V, 309 pages. 1984.

Vol. 1053: P. Hilton, Nilpotente Gruppen und nilpotente Räume. V, 221 pages. 1984.

Vol. 1054: V. Thomée, Galerkin Finite Element Methods for Parabolic Problems. VII, 237 pages. 1984.

Vol. 1055: Quantum Probability and Applications to the Quantum Theory of Irreversible Processes. Proceedings, 1982. Edited by L. Accardi, A. Frigerio and V. Gorini. VI, 411 pages. 1984.

Vol. 1056: Algebraic Geometry. Bucharest 1982. Proceedings, 1982. Edited by L. Bădescu and D. Popescu. VII, 380 pages. 1984.

Vol. 1057: Bifurcation Theory and Applications. Seminar, 1983. Edited by L. Salvadori. VII, 233 pages. 1984.

Vol. 1058: B. Aulbach, Continuous and Discrete Dynamics near Manifolds of Equilibria. IX, 142 pages. 1984.

Vol. 1059: Séminaire de Probabilités XVIII, 1982/83. Proceedings. Edité par J. Azéma et M. Yor. IV, 518 pages. 1984.

Vol. 1060: Topology. Proceedings, 1982. Edited by L. D. Faddeev and A. A. Mal'cev. VI, 389 pages. 1984.

Vol. 1061: Séminaire de Théorie du Potentiel. Paris, No. 7. Proceedings. Directeurs: M. Brelot, G. Choquet et J. Deny. Rédacteurs: F. Hirsch et G. Mokobodzki. IV, 281 pages. 1984.